Robot Brains

Robot Brains

Circuits and Systems for Conscious Machines

Pentti O. Haikonen
Nokia Research Center, Finland

John Wiley & Sons, Ltd

Copyright © 2007   John Wiley & Sons Ltd, The Atrium, Southern Gate, Chichester,
West Sussex PO19 8SQ, England

Telephone (+44) 1243 779777

Email (for orders and customer service enquiries): cs-books@wiley.co.uk
Visit our Home Page on www.wiley.com

All Rights Reserved. No part of this publication may be reproduced, stored in a retrieval system or transmitted in any form or by any means, electronic, mechanical, photocopying, recording, scanning or otherwise, except under the terms of the Copyright, Designs and Patents Act 1988 or under the terms of a licence issued by the Copyright Licensing Agency Ltd, 90 Tottenham Court Road, London W1T 4LP, UK, without the permission in writing of the Publisher. Requests to the Publisher should be addressed to the Permissions Department, John Wiley & Sons Ltd, The Atrium, Southern Gate, Chichester, West Sussex PO19 8SQ, England, or emailed to permreq@wiley.co.uk, or faxed to (+44) 1243 770620.

This publication is designed to provide accurate and authoritative information in regard to the subject matter covered. It is sold on the understanding that the Publisher is not engaged in rendering professional services. If professional advice or other expert assistance is required, the services of a competent professional should be sought.

Other Wiley Editorial Offices

John Wiley & Sons Inc., 111 River Street, Hoboken, NJ 07030, USA

Jossey-Bass, 989 Market Street, San Francisco, CA 94103-1741, USA

Wiley-VCH Verlag GmbH, Boschstr. 12, D-69469 Weinheim, Germany

John Wiley & Sons Australia Ltd, 42 McDougall Street, Milton, Queensland 4064, Australia

John Wiley & Sons (Asia) Pte Ltd, 2 Clementi Loop #02-01, Jin Xing Distripark, Singapore 129809

John Wily & Sons Canada Ltd, 6045 Freemont Blvd, Mississauga, ONT, L5R 4J3

Wiley also publishes its books in a variety of electronic formats. Some content that appears in print may not be available in electronic books.

Anniversary Logo Design: Richard J. Pacifico

Library of Congress Cataloging in Publication Data

Haikonen, Pentti O.
  Robot brains : circuits and systems for conscious machines / Pentti O. Haikonen.
    p. cm.
  Includes bibliographical references and index.
  ISBN 978-0-470-06204-3 (cloth)
  1. Robotics. 2. Conscious automata. I. Title.
  TJ211.15.H33 2007
  629.8′9263—dc22

2007029110

British Library Cataloguing in Publication Data

A catalogue record for this book is available from the British Library

ISBN-13 978-0470-06204-3

Typeset in 11/14pt Times by Integra Software Services Pvt. Ltd, Pondicherry, India
Printed and bound in Great Britain by Antony Rowe Ltd, Chippenham, Wiltshire
This book is printed on acid-free paper responsibly manufactured from sustainable forestry
in which at least two trees are planted for each one used for paper production.

Contents

Preface ix

1 Introduction 1
  1.1 General intelligence and conscious machines 1
  1.2 How to model cognition? 3
  1.3 The approach of this book 6

2 Information, meaning and representation 9
  2.1 Meaning and the nonnumeric brain 9
  2.2 Representation of information by signal vectors 11
    2.2.1 Single signal and distributed signal representations 11
    2.2.2 Representation of graded values 14
    2.2.3 Representation of significance 14
    2.2.4 Continuous versus pulse train signals 15

3 Associative neural networks 17
  3.1 Basic circuits 17
    3.1.1 The associative function 17
    3.1.2 Basic neuron models 17
    3.1.3 The Haikonen associative neuron 19
    3.1.4 Threshold functions 20
    3.1.5 The linear associator 22
  3.2 Nonlinear associators 24
    3.2.1 The nonlinear associative neuron group 24
    3.2.2 Simple binary associator 26
    3.2.3 Associator with continuous weight values 28
    3.2.4 Bipolar binary associator 29
    3.2.5 Hamming distance binary associator 30
    3.2.6 Enhanced Hamming distance binary associator 30
    3.2.7 Enhanced simple binary associator 31
  3.3 Interference in the association of signals and vectors 32
  3.4 Recognition and classification by the associative neuron group 35
  3.5 Learning 38
    3.5.1 Instant Hebbian learning 38
    3.5.2 Correlative Hebbian learning 39
  3.6 Match, mismatch and novelty 41
  3.7 The associative neuron group and noncomputable functions 42

4 Circuit assemblies — 45
4.1 The associative neuron group — 45
4.2 The inhibit neuron group — 46
4.3 Voltage-to-single signal (V/SS) conversion — 46
4.4 Single signal-to-voltage (SS/V) conversion — 48
4.5 The 'Winner-Takes-All' (WTA) circuit — 49
4.6 The 'Accept-and-Hold' (AH) circuit — 51
4.7 Synaptic partitioning — 52
4.8 Serial-to-parallel transformation — 54
4.9 Parallel-to-serial transformation — 56
4.10 Associative Predictors and Sequencers — 57
4.11 Timing circuits — 61
4.12 Timed sequence circuits — 63
4.13 Change direction detection — 66

5 Machine perception — 69
5.1 General principles — 69
5.2 Perception and recognition — 70
5.3 Sensors and preprocesses — 71
5.4 Perception circuits; the perception/response feedback loop — 72
    5.4.1 The perception of a single feature — 72
    5.4.2 The dynamic behaviour of the perception/response feedback loop — 74
    5.4.3 Selection of signals — 76
    5.4.4 Perception/response feedback loops for vectors — 77
    5.4.5 The perception/response feedback loop as predictor — 79
5.5 Kinesthetic perception — 81
5.6 Haptic perception — 82
5.7 Visual perception — 84
    5.7.1 Seeing the world out there — 84
    5.7.2 Visual preprocessing — 85
    5.7.3 Visual attention and gaze direction — 86
    5.7.4 Gaze direction and visual memory — 87
    5.7.5 Object recognition — 89
    5.7.6 Object size estimation — 91
    5.7.7 Object distance estimation — 92
    5.7.8 Visual change detection — 94
    5.7.9 Motion detection — 95
5.8 Auditory perception — 98
    5.8.1 Perceiving auditory scenes — 98
    5.8.2 The perception of separate sounds — 99
    5.8.3 Temporal sound pattern recognition — 101
    5.8.4 Speech recognition — 102
    5.8.5 Sound direction perception — 103
    5.8.6 Sound direction detectors — 105
    5.8.7 Auditory motion detection — 111
5.9 Direction sensing — 111
5.10 Creation of mental scenes and maps — 113

6	**Motor actions for robots**	**117**
	6.1 Sensorimotor coordination	117
	6.2 Basic motor control	117
	6.3 Hierarchical associative control	120
	6.4 Gaze direction control	122
	6.5 Tracking gaze with a robotic arm	126
	6.6 Learning motor action sequences	128
	6.7 Delayed learning	129
	6.8 Moving towards the gaze direction	129
	6.9 Task execution	131
	6.10 The quest for cognitive robots	134
7	**Machine cognition**	**137**
	7.1 Perception, cognition, understanding and models	137
	7.2 Attention	139
	7.3 Making memories	140
	7.3.1 Types of memories	140
	7.3.2 Short-term memories	140
	7.3.3 Long-term memories	142
	7.4 The perception of time	143
	7.5 Imagination and planning	145
	7.6 Deduction and reasoning	146
8	**Machine emotions**	**149**
	8.1 Introduction	149
	8.2 Emotional significance	150
	8.3 Pain and pleasure as system reactions	150
	8.4 Operation of the emotional soundtrack	152
	8.5 Emotional decision making	153
	8.6 The system reactions theory of emotions	154
	8.6.1 Representational and nonrepresentational modes of operation	154
	8.6.2 Emotions as combinations of system reactions	155
	8.6.3 The external expressions of emotions	156
	8.7 Machine Motivation and willed actions	156
9	**Natural language in robot brains**	**159**
	9.1 Machine understanding of language	159
	9.2 The representation of words	161
	9.3 Speech acquisition	161
	9.4 The multimodal model of language	163
	9.4.1 Overview	163
	9.4.2 Vertical grounding of word meaning	165
	9.4.3 Horizontal grounding; syntactic sentence comprehension	168
	9.4.4 Combined horizontal and vertical grounding	172
	9.4.5 Situation models	173
	9.4.6 Pronouns in situation models	175
	9.5 Inner speech	176

10 A cognitive architecture for robot brains — 179
10.1 The requirements for cognitive architectures — 179
10.2 The Haikonen architecture for robot brains — 180
10.3 On hardware requirements — 183

11 Machine consciousness — 185
11.1 Consciousness in the machine — 185
    11.1.1 The immateriality of mind — 185
    11.1.2 The reportability aspect of consciousness — 186
    11.1.3 Consciousness as internal interaction — 188
11.2 Machine perception and qualia — 190
11.3 Machine self-consciousness — 191
    11.3.1 The self as the body — 191
    11.3.2 The experiencing self — 192
    11.3.3 Inner speech and consciousness — 193
    11.3.4 The continuum of the existence of the self — 194
11.4 Conscious machines and free will — 194
11.5 The ultimate test for machine consciousness — 195
11.6 Legal and moral questions — 198

Epilogue — 201
The dawn of real machine cognition — 201

References — 203

Index — 209

Preface

Should robots understand what they are doing? What is understanding anyhow and could it be realized artificially? These questions relate to the emerging field of machine cognition and consciousness. This is a rather new discipline that is transforming philosophical musings into engineering solutions.

What is machine consciousness? Chrisley, Clowes and Torrance (2005) identify the goals of machine consciousness research as the creation of artefacts that have characteristics typically associated with awareness, self-awareness, emotion, experience, imagination, etc., and the modelling of these in embodied systems such as robots.

The first serious papers about machine consciousness began to appear in the beginning of the 1990s. Taylor (1992, 1997, 1999), Aleksander and Morton (1993) and Aleksander (1996, 2005) began to propose neural mechanisms for consciousness and mind. Duch (1994, 2005) studied a combined neural/symbolic approach to artificial minds. Trehub (1991) and Valiant (1994) presented some algorithms for cognitive processing, but did not address directly the question of machine consciousness. Sloman (2000) outlined the requirements for minds on a more general level. Philosophical approaches were presented, for example, by Dennett (1991) and Baars (1997). Also the author began his work on machine cognition in the same decade and has summarized the foundations of his approach in the book *The Cognitive Approach to Conscious Machines* (Imprint Academic, UK, 2003a).

The common approach to machine cognition has been the traditional program-based way. However, it is known that the ultimate model for cognition, the human brain, is not program based and is definitely not a digital computer. Therefore, there must be another way to realize machine cognition, one that does not depend on programmed computers.

The author visions autonomous robots that perceive and understand the world and act in it in a natural way, without programs and numerical representation of information. This approach considers the cognitive machine as a system that is seamlessly interactive, both internally and externally, in respect to its environment and experience. This approach should result in robots that know and understand what they are doing and why, robots that can plan and imagine their actions and the possible outcome of these. Robots that exhibit properties like these are said to possess machine consciousness, which may or may not have common deeper properties with animal and human consciousness. The creation of these kinds of machines is also expected to illuminate the nature of human consciousness.

This book considers the engineering aspects of nonnumeric cognitive and potentially conscious machines and should be of interest to anyone who wishes to consider the actual realization of these machines and robots. The principles are presented with the aim of producing dedicated circuits and systems, but obviously everything can be simulated; those who prefer programmed simulations should find it rather easy to implement these principles in the way of computer programs.

I thank the head of the Nokia Research Center, Dr Bob Iannucci, and my superiors, Dr Juha Lehikoinen and Dr Kari Laurila, for the possibility of working in this interesting field. I want to express my special gratitude to my older brother Dr Terho (Max) Haikonen for reading the manuscript and giving valuable advice. I wish to thank Dr Markku Åberg and Mr Arto Rantala of the Technical Research Centre of Finland, Centre for Microelectronics, who have also perused the manuscript and have given their expert advice. I also wish to thank Mr Tuomo Alamäki for his help and support.

Extra thanks go to my son Pete for the science fiction underground-style comic strip in the last chapter. Finally, my very special thanks go to my ever-young wife Sinikka for her support and patience.

Pentti O. Haikonen

1
Introduction

1.1 GENERAL INTELLIGENCE AND CONSCIOUS MACHINES

Suppose that you were going to see a conscious machine, perhaps a robot. What would you expect to see? Recent advances in robotics have produced animal and humanoid robots that are able to move in very naturalistic ways. Would you find these robots conscious? I do not think so. As impressive as the antics of these artefacts are, their shortcoming is easy to see; the lights may be on, but there is 'nobody' at home. The program-controlled microprocessors of these robots do not have the faintest trace of consciousness and the robots themselves do not know what they are doing. These robots are no more aware of their own existence than a cuckoo clock on a good day.

Artificial intelligence (AI) has brought chess-playing computer programs that can beat grand masters and other 'intelligent' programs that can execute given specific tasks. However, the intelligence of these programs is not that of the machine; instead it is the intelligence of the programmer who has laid down the rules for the execution of the task. At best and with some goodwill these cases of artificial intelligence might be called 'specific intelligence' as they work only for their specific and limited application. In contrast, 'general intelligence' would be flexible and applicable over a large number of different problems. Unfortunately, artificial general intelligence has been elusive. Machines do not really understand anything, as they do not utilize meanings.

Robots do not fare well in everyday tasks that humans find easy. Artificial intelligence has not been able to create general intelligence and common sense. Present-day robots are definitely not sentient entities. Many researchers have recognized this and see this as a shortcoming that must be remedied if robots are ever to be as versatile as humans. Robots should be conscious.

Exactly which features and abilities would distinguish a conscious robot from its nonconscious counterpart? It may well be that no matter how 'conscious' a robot might be, some philosophers would still pretend to find one or another successful argument against its consciousness.

An engineering approach may bypass the philosophical mess by defining what the concept of 'machine consciousness' would involve. However, care should be taken here to consider first what natural consciousness is like so that the cautionary example of artificial intelligence is not repeated; AI is definitely artificial but has somehow managed to exclude intelligence.

Folk psychology describes human consciousness as 'the immaterial feeling of being here'. This is accompanied by the awareness of self, surroundings, personal past, present and expected future, awareness of pain and pleasure, awareness of one's thoughts and mental content. Consciousness is also linked to thinking and imagination, which themselves are often equated to the flow of inner speech and inner imagery. Consciousness is related to self, mind and free will. Consciousness is also seen to allow one to act and execute motions fluently, without any apparent calculations. A seen object can be readily grasped and manipulated. The environment is seen as possibilities for action.

Folk psychology is not science. Thus it is not able to determine whether the above phenomena were caused by consciousness or whether consciousness is the collection of these phenomena or whether these phenomena were even real or having anything to do with consciousness at all. Unfortunately philosophy, while having done much more, has not done much better.

A robot that could understand the world and perceive it as possibilities for action would be a great improvement over existing robots. Likewise, a robot that would communicate with natural language and would even have the flow of natural language inner speech could easily cooperate with humans; it would be like one of us. It can be seen that the machine implementation of the folk psychology hallmarks of consciousness would be beneficial regardless of their actual relationship to 'true' consciousness, whatever that might eventually turn out to be.

This can be the starting point for the engineering approach. The folk psychology hallmarks of consciousness should be investigated in terms of cognitive sciences and defined in engineering terms for machine implementation. This would lead to an implementable specification for a 'conscious machine'.

This 'conscious machine' would be equipped with sensors and perception processes as well as some means for a physical response. It would perceive the world in a direct way, as objects and properties out there, just as humans do, and perceive itself as being in the centre of the rest of the world. It would be able to process information, to think, in ways seemingly similar to human thinking, and it would have the flow of inner speech and imagery. It would also be able to observe its own thoughts and introspect its mental content. It would perceive this mental content as immaterial. It would motivate itself and have a will of its own. It would judge the world and its own actions by emotional good–bad criteria. It would be aware of its own existence. It would be able to move and execute actions as freely and readily as humans do. It would emulate the processes of the human brain and cognition. It would appear to be conscious.

This leads to two questions: first, how do the human brain and mind actually work and, second, how could the workings of the brain be best emulated in an artificial system? Unfortunately the exact answer is not yet known to the first question and it may well be that a definite answer can be found only after we have learned to create successful artificial minds. Thus, models of machine cognition may also help to model human cognition.

1.2 HOW TO MODEL COGNITION?

Presently there are five main approaches to the modelling of cognition that could be used for the development of cognitive machines: the computational approach (artificial intelligence, AI), the artificial neural networks approach, the dynamical systems approach, the quantum approach and the cognitive approach. Neurobiological approaches exist, but these may be better suited for the eventual explanation of the workings of the biological brain.

The computational approach (also known as artificial intelligence, AI) towards thinking machines was initially worded by Turing (1950). A machine would be thinking if the results of the computation were indistinguishable from the results of human thinking. Later on Newell and Simon (1976) presented their Physical Symbol System Hypothesis, which maintained that general intelligent action can be achieved by a physical symbol system and that this system has all the necessary and sufficient means for this purpose. A physical symbol system was here the computer that operates with symbols (binary words) and attached rules that stipulate which symbols are to follow others. Newell and Simon believed that the computer would be able to reproduce human-like general intelligence, a feat that still remains to be seen. However, they realized that this hypothesis was only an empirical generalization and not a theorem that could be formally proven. Very little in the way of empirical proof for this hypothesis exists even today and in the 1970s the situation was not better. Therefore Newell and Simon pretended to see other kinds of proof that were in those days readily available. They proposed that the principal body of evidence for the symbol system hypothesis was negative evidence, namely the absence of specific competing hypotheses; how else could intelligent activity be accomplished by man or machine? However, the absence of evidence is by no means any evidence of absence. This kind of 'proof by ignorance' is too often available in large quantities, yet it is not a logically valid argument. Nevertheless, this issue has not yet been formally settled in one way or another. Today's positive evidence is that it is possible to create world-class chess-playing programs and these can be called 'artificial intelligence'. The negative evidence is that it appears to be next to impossible to create real general intelligence via preprogrammed commands and computations.

The original computational approach can be criticized for the lack of a cognitive foundation. Some recent approaches have tried to remedy this and consider systems that integrate the processes of perception, reaction, deliberation and reasoning (Franklin, 1995, 2003; Sloman, 2000).

There is another argument against the computational view of the brain. It is known that the human brain is slow, yet it is possible to learn to play tennis and other activities that require instant responses. Computations take time. Tennis playing and the like would call for the fastest computers in existence. How could the slow brain manage this if it were to execute computations?

The artificial neural networks approach, also known as connectionism, had its beginnings in the early 1940s when McCulloch and Pitts (1943) proposed that the brain cells, neurons, could be modelled by a simple electronic circuit. This circuit would receive a number of signals, multiply their intensities by the so-called synaptic weight values and sum these modified values together. The circuit would give an output signal if the sum value exceeded a given threshold. It was realized that these artificial neurons could learn and execute basic logic operations if their synaptic weight values were adjusted properly. If these artificial neurons were realized as hardware circuits then no programs would be necessary and biologically plausible artificial replicas of the brain might be possible. Also, neural networks operate in parallel, doing many things simultaneously. Thus the overall operational speed could be fast even if the individual neurons were slow.

However, problems with artificial neural learning led to complicated statistical learning algorithms, ones that could best be implemented as computer programs. Many of today's artificial neural networks are statistical pattern recognition and classification circuits. Therefore they are rather removed from their original biologically inspired idea. Cognition is not mere classification and the human brain is hardly a computer that executes complicated synaptic weight-adjusting algorithms.

The human brain has some 10^{11} neurons and each neuron may have tens of thousands of synaptic inputs and input weights. Many artificial neural networks learn by tweaking the synaptic weight values against each other when thousands of training examples are presented. Where in the brain would reside the computing process that would execute synaptic weight adjusting algorithms? Where would these algorithms have come from? The evolutionary feasibility of these kinds of algorithms can be seriously doubted. Complicated algorithms do not evolve via trial and error either. Moreover, humans are able to learn with a few examples only, instead of having training sessions with thousands or hundreds of thousands of examples. It is obvious that the mainstream neural networks approach is not a very plausible candidate for machine cognition although the human brain is a neural network.

Dynamical systems were proposed as a model for cognition by Ashby (1952) already in the 1950s and have been developed further by contemporary researchers (for example Thelen and Smith, 1994; Gelder, 1998, 1999; Port, 2000; Wallace, 2005). According to this approach the brain is considered as a complex system with dynamical interactions with its environment. Gelder and Port (1995) define a dynamical system as a set of quantitative variables, which change simultaneously and interdependently over quantitative time in accordance with some set of equations. Obviously the brain is indeed a large system of neuron activity variables that change over time. Accordingly the brain can be modelled as a dynamical system if the neuron activity can be quantified and if a suitable set of, say, differential equations can be formulated. The dynamical hypothesis sees the brain as comparable to analog feedback control systems with continuous parameter values. No inner representations are assumed or even accepted. However, the dynamical systems approach seems to have problems in explaining phenomena like 'inner speech'. A would-be designer of an artificial brain would find it difficult to see what kind of system dynamics would be necessary for a specific linguistically expressed thought. The dynamical systems approach has been criticized, for instance by Eliasmith (1996, 1997), who argues that the low dimensional systems of differential equations, which must rely on collective parameters, do not model cognition easily and the dynamicists have a difficult time keeping arbitrariness from permeating their models. Eliasmith laments that there seems to be no clear ways of justifying parameter settings, choosing equations, interpreting data or creating system boundaries. Furthermore, the collective parameter models make the interpretation of the dynamic system's behaviour difficult, as it is not easy to see or determine the meaning of any particular parameter in the model. Obviously these issues would translate into engineering problems for a designer of dynamical systems.

The quantum approach maintains that the brain is ultimately governed by quantum processes, which execute nonalgorithmic computations or act as a mediator between the brain and an assumed more-or-less immaterial 'self' or even 'conscious energy field' (for example Herbert, 1993; Hameroff, 1994; Penrose, 1989; Eccles, 1994). The quantum approach is supposed to solve problems like the apparently nonalgorithmic nature of thought, free will, the coherence of conscious experience, telepathy, telekinesis, the immortality of the soul and others. From an engineering point of view even the most practical propositions of the quantum approach are presently highly impractical in terms of actual implementation. Then there are some proposals that are hardly distinguishable from wishful fabrications of fairy tales. Here the quantum approach is not pursued.

The cognitive approach maintains that conscious machines can be built because one example already exists, namely the human brain. Therefore a cognitive machine should emulate the cognitive processes of the brain and mind, instead of merely trying to reproduce the results of the thinking processes. Accordingly the results of neurosciences and cognitive psychology should be evaluated and implemented in the design if deemed essential. However, this approach does not necessarily involve the simulation or emulation of the biological neuron as such, instead, what is to be produced is the abstracted information processing function of the neuron.

A cognitive machine would be an embodied physical entity that would interact with the environment. Cognitive robots would be obvious applications of machine cognition and there have been some early attempts towards that direction. Holland seeks to provide robots with some kind of consciousness via internal models (Holland and Goodman, 2003; Holland, 2004). Kawamura has been developing a cognitive robot with a sense of self (Kawamura, 2005; Kawamura *et al.*, 2005). There are also others. Grand presents an experimentalist's approach towards cognitive robots in his book (Grand, 2003).

A cognitive machine would be a complete system with processes like perception, attention, inner speech, imagination, emotions as well as pain and pleasure. Various technical approaches can be envisioned, namely indirect ones with programs, hybrid systems that combine programs and neural networks, and direct ones that are based on dedicated neural cognitive architectures. The operation of these dedicated neural cognitive architectures would combine neural, symbolic and dynamic elements. However, the neural elements here would not be those of the traditional neural networks; no statistical learning with thousands of examples would be implied, no backpropagation or other weight-adjusting algorithms are used. Instead the networks would be associative in a way that allows the symbolic use of the neural signal arrays (vectors). The 'symbolic' here does not refer to the meaning-free symbol manipulation system of AI; instead it refers to the human way of using symbols with meanings. It is assumed that these cognitive machines would eventually be conscious, or at least they would reproduce most of the folk psychology hallmarks of consciousness (Haikonen, 2003a, 2005a). The engineering aspects of the direct cognitive approach are pursued in this book.

1.3 THE APPROACH OF THIS BOOK

This book outlines an engineering approach towards cognitive and conscious machines. These machines would perceive the world directly, as objects and properties out there, and integrate these percepts seamlessly into their cognitive processes. They would have the flow of inner speech and imagery. They would observe and introspect their mental content and perceive this mental content as immaterial. They would judge the world and their own actions by emotional good–bad criteria. They would be self-motivated agents with a will of their own. They would be able to move and execute actions as freely and readily as humans do. They would produce the hallmarks of consciousness.

The requirements for the direct perception of the world and the seamless integration would seem to call for a special way of perception and information representation. Here the traditional way of information representation of the artificial intelligence, the representation of information as statements in a formal language, is rejected. Instead, associative nonnumeric neural networks, distributed representations and cognitive architectures are investigated as the solution.

The book chapters proceed in logical order and build on previously presented matter; thus they should be studied sequentially. Some matters may appear trivial to advanced readers. However, in that case the reader should pay attention as apparently familiar matters may also have novel twists and meanings in this context. The last chapter, 'Machine Consciousness', should also be accessible to enlightened nonengineer readers who are interested in machine consciousness research. The contents of the book chapters are summarized here.

Chapter 1, 'Introduction', is the introduction in the present chapter.

Chapter 2, 'Information, Meaning and Representation', notes that humans operate with meanings while computers normally do not. It is argued that the brain is not a digital computer and therefore a cognitive robot should not be one either. Instead of digital representation of information a nonnumeric information representation by large signal arrays, vectors, is proposed. This method is used in the approach of this book.

Chapter 3, 'Associative Neural Networks', presents neural networks that can be used for the associative processing of signal vectors. With these networks vectors can be associated with each other and can be evoked by each other. The neural network that executes this operation is called an associator. The limitations of the traditional neural associator are pointed out and enhanced associators that remedy these limitations are introduced. These associators are based on artificial associative neurons and here one specific execution of these, the Haikonen associative neuron, is presented. The proposed associators are not statistical neural networks and do not utilize learning algorithms such as 'backpropagation'.

Chapter 4, 'Circuit Assemblies', presents some example realizations and definitions for the basic circuit assemblies that are used as building blocks in the systems that are introduced later on in this book. These circuits include the basic associative neuron group, a 'Winner-Takes-All' circuit, an 'Accept-and-Hold' circuit, associative predictors and sequencers, timed sequence circuits and others.

Chapter 5, 'Machine Perception', describes the principles and circuits that would allow direct perception of the world, apparently as such in the machine. The associative neuron-based perception/response feedback loop is introduced as the fundamental building block that is applicable to every sensory modality, such as kinesthetic, haptic, visual and auditory perception. The purpose of this chapter is not to give an exhaustive treatment on the issues of perception, which would be many; instead the purpose is to present the basic ideas that are necessary for the understanding of the cognitive system that is outlined in the following chapters.

Chapter 6, 'Motor Actions for Robots', describes the integration of motor systems into the associative system. According to this approach a robot will be able to execute motor actions in the directly perceived environment without any numeric computations. Hierarchical control loops allow the planning and evocation of action in associative terms and the control of the actual execution in motor control terms. Simple control examples are described.

Chapter 7, 'Machine Cognition', argues that reactive processes are not sufficient for an autonomous robot and deliberative processes are needed. These processes call for higher-level cognition and the use of mental models. The issues of understanding, memories, perception of time, imagination and reasoning in the cognitive machine are considered.

Chapter 8, 'Machine Emotions', describes how the functional equivalents of pain, pleasure, good and bad can be evolved into a system of values, motivations, attention and learning control, and emotional decision making. The Haikonen System Reactions Theory of Emotions is applied to the cognitive machine.

Chapter 9, 'Natural Language in Robot Brains', outlines how natural language can be used and understood by the machine using the associative neural networks. The author's 'multimodal model of language' that seamlessly integrates language, multisensory perception and motor actions is utilized here. The grounding of word meaning is explained. The robot utilizes natural language also as 'inner speech', which is a running self-commentary of the moment-to-moment situation of the robot.

Chapter 10, 'A Cognitive Architecture for Robot Brains', summarizes the presented principles in an outline for the system architecture of the complete cognitive machine, the 'robot brain'. This machine is not a digital computer; instead it is a system based on distributed signal representations that are processed by associative neuron groups and additional circuits. This system is embodied, it has sensors and motor systems and it utilizes meanings that are grounded to the percepts of environment and the system itself, both the body and the 'mental content'. This system is also self-motivated with values.

Chapter 11, 'Machine Consciousness', summarizes the essential issues of machine consciousness and explains how the approach of this book may produce machines that can be called conscious. The immateriality of mind, the reportability aspect of consciousness, the internal interaction aspect of consciousness, qualia, self-consciousness, free will, testing of consciousness and other issues are considered. It is argued that the presented approach can produce machines that exhibit most, if not all, of the folk psychology hallmarks of consciousness. The chapter concludes with some legal and moral issues that may arise when cognitive robots are introduced.

Please note that the circuit diagrams and other construction examples in this book are presented only for illustrative purposes; they are not intended to be taken as actual product construction guidance. Anyone trying to replicate these circuits and systems should consider the safety and other requirements of the specific application.

2
Information, meaning and representation

2.1 MEANING AND THE NONNUMERIC BRAIN

Humans perceive the world in a seemingly direct way. They see the world and its objects to be out there and hear the sounds of the world and understand the meanings of them. Each sensory modality represents the sensed information directly, as the actual qualities of the sensed entities – or so it seems. However, it is known today that this appearance arises via rather complicated sensory and cognitive processes in the brain. These processes are carried out by neural activities that remain beyond awareness. Humans cannot perceive the material structure of the brain, nor the neural firings that take place there. Somehow all this remains transparent, but what is perceived is the actual information without any apparent material base. Obviously a true cognitive machine should experience the information in a similar way. Therefore the following questions must be asked. Which kinds of methods, signals and systems could carry and process meaning in a way that would appear transparent and seemingly immaterial to the robot brain? Would these methods differ from the information representation methods that are used in present-day digital computers?

What is information in a computer? It is understood that all information may be digitized and represented by patterns of bits, ones and zeros to almost any degree of accuracy. Digital technology has its roots in Shannon's information theory (Shannon, 1948). Shannon's mathematical theory of communication arose from the requirement of faithful information transmission. Shannon noted that the fundamental problem of communication related to the errorless reproduction of the transmitted message at the receiving end. In order to solve this Shannon had to give an exact definition to 'information'. While doing this he noticed that the actual meaning of the message was not relevant to this problem at all. Thus Shannon's definition of information came to relate only to the transmitted bit patterns and not to any possible meanings that may be carried by these bits. Shannon's theory has proved to be extremely useful in digital communication engineering, where the task is to transmit and receive messages without distortion through noisy bandwidth-limited channels. In that context the actual meaning of the message is not considered. A mobile phone user can be confident that the message is carried over in the same way regardless of the topic of the talk. The actual meaning is nevertheless carried by the electronic signals, but it remains to be decoded by the listener.

In humans various sensors like the eyes and ears transmit information to the brain. This transmission takes place via signals that travel through neural fibres. Thus, should senses be considered as encoders and transmitters of information and the brain as the receiver of information in Shannon's sense? It could, but this approach might complicate the aspects of meaning. Where does the meaning come from? Who interprets the meaning if the brain is only a receiver that tries to reproduce the transmitted signal pattern? Who watches the received signal patterns and assigns meaning to them? This approach can easily lead to the homunculus model of the mind. In this failed model the sensory information is reassembled and displayed at an internal 'Cartesian theater' while the homunculus, the 'small man inside', the self, is watching the show and understanding the meaning.

Therefore, here is the difference. Shannon's information theory does not consider the actual meaning of messages. Also, the present-day computer tries to do without meanings; it does not understand the meanings of the computations that it executes. A word processing program does not understand the text that is being typed in. The symbolic processing with binary words and syntax is inherently meaning-free. On the other hand, human cognition is about meanings. Humans perceive sensory information directly as objects, entities and qualities without any apparent need to observe and decode neural signal patterns. The association of meaning with perceived entities is automatic and unavoidable. Visual patterns are perceived as things with possibilities for action as easily and immediately as the meanings of letters and words. Once some one has learned to read, the meanings of words reveal themselves immediately and there is no easy way to stare at letters and words without understanding what they mean. However, the material neural apparatus that executes this remains transparent and thoughts appear seemingly immaterial to us.

This is the grand challenge of robot brain design. Which kind of information representation method could carry and process meaning in a way that would appear transparent and seemingly immaterial to the robot brain? Which kind of information representation method would allow imagination in a robot brain? Which kind of information representation method would allow symbolic thought? The traditional way has been to digitize everything, represent everything with numbers, use digital signal processing and invent clever algorithms. However, the brain is hardly a digital calculator that utilizes complicated digital signal processing algorithms. Therefore, there must be another way, a way of information representation with inherent meaning and without numeric values. This kind of direct and *nonnumeric* way is pursued in the following.

2.2 REPRESENTATION OF INFORMATION BY SIGNAL VECTORS

2.2.1 Single signal and distributed signal representations

In principle, real world objects may be described and identified by their typical properties. For example, a cherry may be described as a small red ball that is soft and tastes sweet and grows on trees. Consequently, a cherry could be identified by detecting the presence of these properties. Therefore a hypothetical cherry-detecting machine should have specific detectors for at least some of these properties. As a first approximation on/off detections would suffice; a property is either present or absent. On the other hand, on/off information can be expressed by *ones* and *zeros*, as the presence or absence of a signal. This would lead to the representation by signal vectors, arrays of *ones* and *zeros*, where each *one* would indicate the presence of the corresponding property or feature.

According to Figure 2.1, a cherry could be represented by a signal vector 100 100 100. This representation is better than the name 'cherry' as it tells something about the actual appearance of the cherry provided that the meanings of the individual signals are known. This would take place if these meanings were grounded to the detectors that detect their specific attributes or features. This is the basic idea behind the representation by signal vectors, also known as distributed representations. Distributed representations were proposed by Hinton, McClelland and Rumelhart in the 1980s (Hinton *et al.*, 1990). In distributed representations each object or entity is represented by an activity that is distributed over a wide range of signal lines and computing nodes. Moreover, these lines and nodes are involved in the representation of many different entities. Distributed representations are usually seen as the opposite to local representations, where each signal represents only one entity. These representations are also known as grandmother representations (as there would also be one specific signal for the representation of your grandmother). Here this kind of strict distinction is not endorsed. Instead, the grandmother representation is seen as a special case of distributed representations. This is called here the 'single signal representation'.

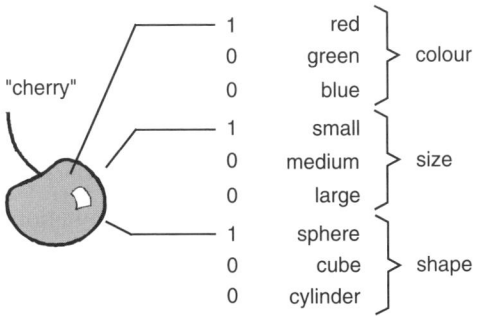

Figure 2.1 Representation of an object with on/off feature signals

12 INFORMATION, MEANING AND REPRESENTATION

Table 2.1 An example of a single signal vector representation: size property

Vector	Property
0 0 1	Small
0 1 0	Medium
1 0 0	Large

Table 2.2 An example of a fully distributed signal representation: colours

Vector	Colour
0 0 1	Blue
0 1 0	Green
0 1 1	Cyan
1 0 0	Red
1 0 1	Magenta
1 1 0	Yellow
1 1 1	White

In the single signal representation each signal is involved only in the representation of one property, feature or entity. Thus signals that originate from feature detectors and depict the presence of one specific feature are necessarily single signal representations. An example of a single signal representation vector is given in Table 2.1.

In Table 2.1 the properties small, medium and large are mutually exclusive. For instance, if an object is seen as small then it cannot be large. When one signal is assigned to each of these properties then a three-signal vector arises and due to the mutual property exclusivity only three different vectors are allowed.

In fully distributed representation the signals are not mutually exclusive; each signal may have the value of one or zero independent of the other signals. As an example, the representation of colours by three primary colour signals is considered in Table 2.2.

The colours in Table 2.2 correspond to the perceived colours that would arise from the combination of the primary colours red, green and blue in a similar way as on the television screen. It can be seen that seven different colours can be represented by three on/off signals. The vector 0 0 0 would correspond to the case where no primary colours were detected – black.

In practice an entity is often represented by combinations of groups of single signal representations. Here the individual groups represent mutually exclusive features, while these groups are not mutually exclusive. This representation therefore combines the single signal representation and the fully distributed representation. As an example, Table 2.3 depicts the representation of a three-letter word in this way.

Table 2.3 An example of the mixed signal representation: the word 'cat'

First letter a b c ... t ... x y z	Second letter a b c ... t ... x y z	Third letter a b c ... t ... x y z
0 0 1 ... 0 ... 0 0 0	1 0 0 ... 0 ... 0 0 0	0 0 0 ... 1 ... 0 0 0

It is obvious that in the representation of a word the first, second, third, etc., letter may be one and only one of the alphabet, while being independent of the other letters. Thus words would be represented by m signal groups, each having n signals, where m is the number of letters in the word and n is the number of allowable alphabets.

These three different cases of signal vector representations have different representational capacities per signal line. The maximum number of different vectors with n signals is 2^n; hence in theory this is the maximum representational capacity of n signal lines. However, here the signal value zero represents the absence of its corresponding feature and therefore all-zero vectors represent nothing. Thus the representational capacity of a fully distributed representation will be $2^n - 1$. The properties of the single signal, mixed, fully distributed and binary representations are summarized in Table 2.4.

It can be seen that the single signal and mixed representations are not very efficient in their use of signal lines. Therefore the system designer should consider carefully when to utilize each of these representations.

Signal vector representations are large arrays of ones and zeros and they do not superficially differ from equal length binary numbers. However, their meaning is different. They do not have a numerical meaning; instead their meaning is grounded to the system that carries the corresponding signals and eventually to the specific feature detectors. Therefore this meaning is not portable from one system to another unless the systems have exactly similar wiring. On the other hand, binary numbers have a universal meaning that is also portable.

Mixed and fully distributed representations have a fine structure. The appearance of the represented entity may be modified by changing some of the constituent features by changing the ones and zeros. In this way an entity may be made smaller

Table 2.4 Summary of the single signal, mixed, fully distributed and binary vector representations

Vector type	Example	Number of signal lines		Number of possibilities	
Single signal	001000	n	(6)	n	(6)
Mixed	100001	$n \times m$	(3×2)	n^m	(9)
Fully distributed	101001	n	(6)	$2^n - 1$	(63)
Binary	101001	n	(6)	2^n	(64)

14 INFORMATION, MEANING AND REPRESENTATION

or bigger, a different colour, a different posture, etc. This is a useful property for machine reasoning and imagination.

2.2.2 Representation of graded values

In the previous treatment a property or feature was assumed to be present or not present and thus be presentable by one or zero, the presence or absence of the corresponding signal. However, properties may have graded values. For instance, objects may have different shades of colours and greyness, from light to dark. In these cases the gradation may be represented by single signal vectors, as shown in Table 2.5. Other graded values, such as the volume of a sound, may be represented in a similar way.

2.2.3 Representation of significance

The meaning of a signal vector representation is carried by the presence of the individual signals; therefore the actual signal intensity does not matter. Thus the multiplication of the signal intensity by a coefficient will not change the meaning of the signal, because the meaning is hardwired to the point of origination:

$$\text{meaning } \{s\} = \text{meaning } \{k*s\} \qquad (2.1)$$

where

$s =$ signal intensity of the signal s
$k =$ coefficient

This means that the signal intensity is an additional variable, which is thus available for other purposes. The signal intensity may be modulated by the instantaneous significance of the signal, and simple threshold circuits may be used to separate important and less important signals from each other.

The signal intensity may also be a measure of the confidence of the originator, such as a feature detector, of the signal. A feature detector that detects its dedicated feature with great confidence may output its signal at a high level, while another

Table 2.5 The representation of the graded grey scale

Gradation	Vector
Light	0 0 0 0 1
Light–medium	0 0 0 1 0
Medium	0 0 1 0 0
Medium–dark	0 1 0 0 0
Dark	1 0 0 0 0

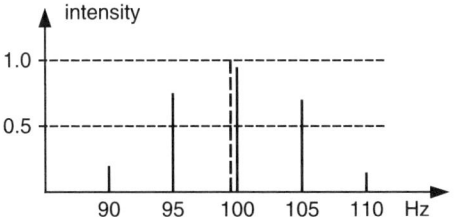

Figure 2.2 Signal frequency detector outputs (a filter bank)

feature detector may output its signal at a lower level. As an example consider a number of filters that are tuned to detect the frequencies 90, 95, 100, 105 and 110 Hz (Figure 2.2).

In Figure 2.2 the actual input frequency is 99 Hz. Consequently the 100 Hz filter has the highest output, but due to the limited filter selectivity other filters also output a signal, albeit at lower levels. A threshold circuit can be used to select the 100 Hz filter output signal as the final result.

2.2.4 Continuous versus pulse train signals

Here continuous signals have been considered. However, it is known that in the brain pulse train signals are used; these pulses have constant amplitude and width while their instantaneous repetition frequency varies. It is possible to use pulse train signals also in artificial circuits with signal vector representations that indicate the presence and significance of the associated property. It can be shown that the pulse repetition frequency of a pulse train signal and the intensity of a continuous signal can be interchanged. Consider Figure 2.3.

The average level of the pulse train signal of Figure 2.3 can be computed as follows:

$$U_a = U * \tau / T \quad (2.2)$$

where

U_a = average level
U = pulse amplitude (constant)
T = period (seconds)
τ = pulse width (seconds) (constant)

The pulse repetition frequency f is

$$f = 1/T \quad (2.3)$$

From Equations (2.2) and (2.3),

$$U_a = U * \tau * f \quad (2.4)$$

16 INFORMATION, MEANING AND REPRESENTATION

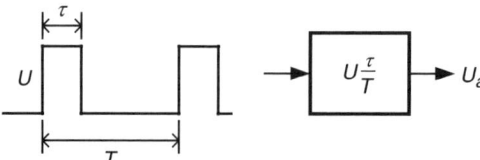

Figure 2.3 The pulse train signal and the averaging circuit

Thus the average level U_a is directly proportional to the pulse repetition frequency f. Consequently, the repetition frequency may be set to carry the same information as the level of a continuous signal. Circuits that operate with continuous signals can be made to accept pulse train signals when the inputs of these circuits are equipped with averaging filters that execute the operation of Equation (2.2). Pulse position modulation may be utilized to carry additional information, but that possibility is not investigated here. In the following chapters continuous signals are assumed.

3
Associative neural networks

3.1 BASIC CIRCUITS

3.1.1 The associative function

Association is one of the basic mechanisms of cognition. Association connects two entities with each other so that one of these entities may be evoked by the other one. The entities to be associated with each other may be represented by signals and arrays of signals' signal vectors. An algorithm or a device that associates signals or signal vectors with each other is called an associator. An associative memory associates two vectors with each other so that the presentation of the first vector will evoke the second vector. In an autoassociative memory the evoking vector is a part of the evoked vector. In a heteroassociative memory the associated vectors are arbitrary. 'Associative learning' refers to mechanisms and algorithms that execute association automatically when certain criteria are met. In the following, artificial neurons and neuron groups for the association of signal vectors are considered.

3.1.2 Basic neuron models

The McCulloch–Pitts neuron (McCulloch and Pitts, 1943) is generally considered as the historical starting point for artificial neural networks. The McCulloch–Pitts neuron is a computational unit that accepts a number of signals $x(i)$ as inputs, multiplies each of these with a corresponding weight value $w(i)$ and sums these products together. This sum value is then compared to a threshold value and an output signal y is generated if the sum value exceeds the threshold value. The McCulloch–Pitts neuron may be depicted in the way shown in Figure 3.1.

Operation of the McCulloch–Pitts neuron can be expressed as follows:

$$\text{IF } \sum w(i)^*x(i) \geq threshold \text{ THEN } y=1 \text{ ELSE } y=0 \qquad (3.1)$$

where

$y=$ output signal
$\sum w(i)^*x(i) =$ evocation sum

Robot Brains: Circuits and Systems for Conscious Machines Pentti O. Haikonen
© 2007 John Wiley & Sons, Ltd

18 ASSOCIATIVE NEURAL NETWORKS

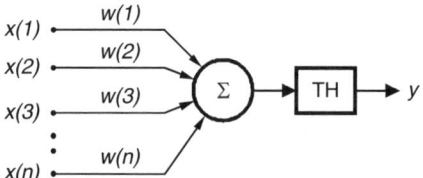

Figure 3.1 The McCulloch–Pitts neuron

$x(i)$ = input signal
$w(i)$ = weight value

The McCulloch–Pitts neuron rule can be reformulated as follows:

$$\text{IF } \Sigma w(i)^*x(i) - threshold \geq 0 \text{ THEN } y=1 \text{ ELSE } y=0 \tag{3.2}$$

The perceptron of Frank Rosenblatt is configured in this way (Rosenblatt, 1958). Here the *threshold* value is taken as the product of an additional fixed input $x(0)=1$ and the corresponding variable weight value $w(0)$. In this way the fixed value of zero may be used as the output threshold. The neuron rule may be rewritten as:

$$\text{IF} \Sigma w(i)^*x(i) \geq 0 \text{ THEN } y=1 \text{ ELSE } y=0 \tag{3.3}$$

In the rule (3.3) the term $w(0)^*x(0)$ has a negative value that corresponds to the desired threshold. The term $w(0)^*x(0)$ is also called 'the bias'. The perceptron is depicted in Figure 3.2.

The main applications of the McCulloch–Pitts neuron and the perceptron are pattern recognition and classification. Here the task is to find the proper values for the weights $w(i)$ so that the output threshold is exceeded when and only when the desired input vector or desired set of input vectors $\{x(1), x(2), \ldots, x(m)\}$ is presented to the neuron. Various algorithms for the determination of the weight values exist. The performance of these neurons depends also on the allowable range of the input and weight values. Are positive and negative values accepted, are continuous values accepted or are only binary values of one and zero accepted? In the following these issues are considered in the context of associators.

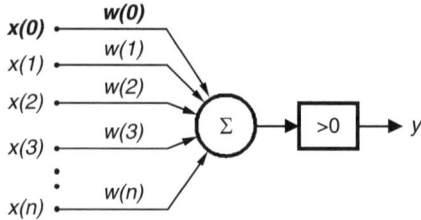

Figure 3.2 The perceptron of Frank Rosenblatt

3.1.3 The Haikonen associative neuron

The Haikonen associative neuron (Haikonen, 1999a, 2003b) is especially devised to associate a signal vector with one signal, the so-called main signal. This neuron utilizes modified correlative Hebbian learning with binary valued (zero or one) synaptic weights. The neuron has also match (m), mismatch (mm) and novelty (n) detection.

In Figure 3.3 s is the so-called main signal, so is the output signal, sa is the associatively evoked output signal and the signals $\{a(1), a(2), \ldots, a(m)\}$ constitute the associative input signal vector A. The number of synapses in this neuron is m and the corresponding synaptic weights are $w(1), w(2), \ldots, w(m)$. The switch SW is open or closed depending on the specific application of the neuron. The output so depends on the state of the switch SW:

$$so = sa \quad \text{when the switch } SW \text{ is open}$$
$$so = s + sa \quad \text{when the switch } SW \text{ is closed}$$

The associatively evoked output signal is determined as follows:

$$\text{IF } \Sigma\,(w(i) \blacklozenge a(i)) \geq threshold \text{ THEN } sa = 1 \text{ ELSE } sa = 0 \quad (3.4)$$

where $\blacklozenge$ is a computational operation (e.g. multiplication).

Match, mismatch and novelty condition detection is required for various operations, as will be seen later. Neuron level match, mismatch and novelty states arise from the instantaneous relationship between the input signal s and the associatively evoked output signal sa. The match m, mismatch mm and novelty n signals are determined as follows:

$$m = s \text{ AND } sa \quad (3.5)$$
$$mm = (\text{NOT } s) \text{ AND } sa \quad (3.6)$$
$$n = s \text{ AND } (\text{NOT } sa) \quad (3.7)$$

where s and sa are rounded to have the logical values of 0 or 1 only.

The match condition occurs when the signal s and the associatively evoked output signal sa coincide and mismatch occurs when the sa signal occurs in the absence of

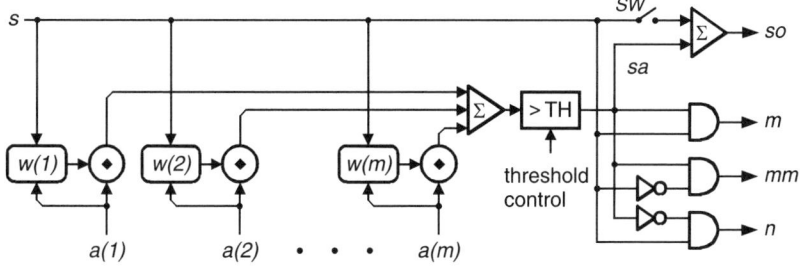

Figure 3.3 The Haikonen associative neuron

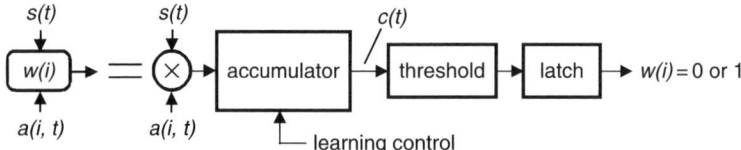

Figure 3.4 The synaptic weight circuit for the Haikonen neuron

the s signal. The novelty condition occurs when the signal s occurs alone or there is no associative connection between a simultaneously active associative input signal vector A and the signal s.

The synaptic weight circuits learn and store the associative connection between an associative signal $a(i)$ and the main signal s. The synaptic weight circuit for the Haikonen neuron is depicted in Figure 3.4.

The synaptic weights $w(i)$ are determined by the correlation of the main signal $s(t)$ and the associative input signal $a(i, t)$. For this purpose the product $s(t)*a(i, t)$ is computed at the moment of learning and the result is forwarded to an accumulator, which stores the so-called correlation sum $c(i, t)$. If the product $s(t)*a(i, t)$ is one then the correlation sum $c(i, t)$ is incremented by a certain step. If the product $s(t)*a(i, t)$ is zero then the correlation sum $c(i, t)$ is decremented by a smaller step. Whenever the correlation sum $c(i, t)$ exceeds the set threshold, the logical value 1 is stored in the latch. The latch output is the synaptic weight value $w(i)$. Instant learning is possible when the threshold is set so low that already the first coincidence of $s(t) = 1$ and $a(i, t) = 1$ drives the correlation sum $c(i, t)$ over the threshold value. A typical learning rule is given below:

$$c(i, t) = c(i, t-1) + 1.5*s(t)*a(i, t) - 0.5*s(t)$$
$$\text{IF } c(i, t) > threshold \text{ THEN } w(i) \Rightarrow 1$$
(3.8)

where

$w(i) =$ synaptic weight, initially $w(i) = 0$

$c(i, t) =$ correlation sum at the moment t

$s(t) =$ input of the associative neuron at the moment t; zero or one

$a(i, t) =$ associative input of the associative neuron at the moment t; zero or one

The association weight value $w(i) = 1$ gained at any moment of association will remain permanent. The rule (3.8) is given here as an example only; variations are possible.

3.1.4 Threshold functions

A threshold circuit compares the intensity of the incoming signal to a threshold value and generates an output value that depends on the result of the

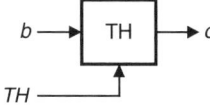

Figure 3.5 A threshold circuit

comparison. Threshold circuits are utilized in various places in associative neurons and networks.

In the threshold circuit of Figure 3.5 b is the input signal that is compared to the threshold level TH and c is the output signal. The threshold level TH may be fixed or may be varied by some external means. There are various possibilities for the threshold operation. The following threshold functions are used in the next chapters.

The linear threshold function circuit has a piecewise linear input–output function. This circuit will output the actual input signal if the intensity of the input signal equals or exceeds the threshold value. The linear threshold function preserves any significance information that may be coded into the intensity of the signal:

$$\text{IF } b < TH \text{ THEN } c = 0$$
$$\text{IF } b \geq TH \text{ THEN } c = b \tag{3.9}$$

The limiting threshold function circuit will output a constant value (logical one) if the intensity of the input signal equals or exceeds the threshold value. The limiting threshold function removes any significance information that may be coded into the intensity of the signal:

$$\text{IF } b < TH \text{ THEN } c = 0$$
$$\text{IF } b \geq TH \text{ THEN } c = 1 \tag{3.10}$$

The linear and limiting threshold functions are presented in Figure 3.6.

The *Winner-Takes-All* threshold can be used to select winning outputs from a group of signals such as the outputs of neuron groups. In this case each signal has its own threshold circuit. These circuits have a common threshold value, which is set to equal or to be just below the maximum value of the intensities of the individual signals. Thus only the signal with the highest intensity will be selected and will generate output. If there are several signals with the same highest intensity then they

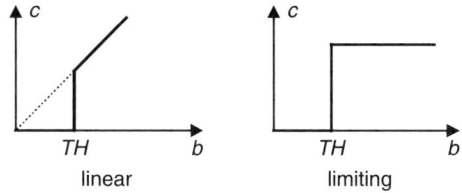

Figure 3.6 Linear and limiting threshold functions

22 ASSOCIATIVE NEURAL NETWORKS

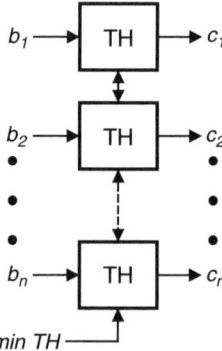

Figure 3.7 The Winner-Takes-All threshold arrangement

will all be selected. The threshold circuit arrangement for the Winner-Takes-All threshold function operation is presented in Figure 3.7.

In Figure 3.7 the input signals are $b_1, b_2, \ldots, b_n$, of which the threshold circuits must select the strongest. The corresponding output signals are $c_1, c_2, \ldots, c_n$.

The Winner-Takes-All threshold may utilize the linear threshold function or the limiting threshold function. A minimum threshold value may be applied to define minimum signal intensities that are allowed to cause output:

$$
\begin{aligned}
&\text{IF } b_i < min\ TH \text{ THEN } c_i = 0 \\
&\text{IF } max\{b\} \geq min\ TH \text{ THEN} \\
&\quad \text{IF } b_i < max\{b\} \text{ THEN } c_i = 0 \\
&\quad \text{IF } b_i = max\{b\} \text{ THEN } c_i = b_i \quad \text{(linear threshold function)}
\end{aligned}
\tag{3.11}
$$

or

$$\text{IF } b_i = max\{b\} \text{ THEN } c_i = 1 \quad \text{(limiting threshold function)}$$

In certain applications a small tolerance may be defined for the $max\{b\}$ threshold value so that signals with intensities close enough to the $max\{b\}$ value will be selected.

3.1.5 The linear associator

The traditional linear associator may be considered as a layer of McCulloch–Pitts neurons without the nonlinear output threshold (see, for instance, Churchland and Sejnowski, 1992, pp. 77–82). Here the task is to associate an output vector $\{y(1), y(2), \ldots, y(m)\}$ with an input vector $\{x(1), x(2), \ldots, x(n)\}$ (see Figure 3.8).

Each neuron has the same input, the $x(j)$ vector. The weight values are different for each neuron; therefore the weight values form a weight matrix $w(i, j)$. The

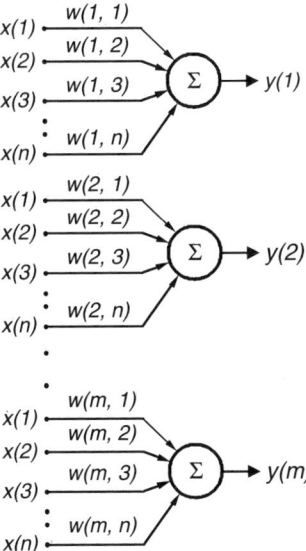

Figure 3.8 The linear associator as a one-layer neural network

output vector $y(i)$ of the linear associator is computed as the inner product of the weight matrix $w(i, j)$ and the input vector $x(j)$ as follows:

$$y(i) = \Sigma w(i, j)^* x(j) \tag{3.12}$$

where the summing index j runs from 1 to n. Equation (3.12) can be expressed in matrix form as

$$\begin{bmatrix} y(1) \\ y(2) \\ y(3) \\ \vdots \\ y(m) \end{bmatrix} = \begin{bmatrix} w(1,1) & w(1,2) & \ldots & w(1,n) \\ w(2,1) & w(2,2) & \ldots & w(2,n) \\ w(3,1) & w(3,2) & \ldots & w(3,n) \\ \vdots & \vdots & & \vdots \\ w(m,1) & w(m,2) & \ldots & w(m,n) \end{bmatrix} \times \begin{bmatrix} x(1) \\ x(2) \\ x(3) \\ \vdots \\ x(n) \end{bmatrix} \tag{3.13}$$

Basically the linear associator is a set of artificial neurons, which do not have a nonlinear output threshold. These neurons share common input signals, which are forwarded to the neurons via weighted connections, 'synapses'. In the literature there are various depictions for the linear associator. Two common depictions are given in Figure 3.9. Both diagrams depict the same thing.

The linear associator executes a function that maps input vectors into output vectors. For the desired mapping the weight matrix $w(i, j)$ must be determined properly. The linear associator has a rather limited pattern storage capacity and is pestered by phenomena that can be described as 'interference', 'spurious responses' and 'filling up early'. Traditionally improvements for the linear associator have been sought by the use of the nonlinear output threshold, improved weight learning algorithms and sparse coding. These methods have solved the problems of the linear

24 ASSOCIATIVE NEURAL NETWORKS

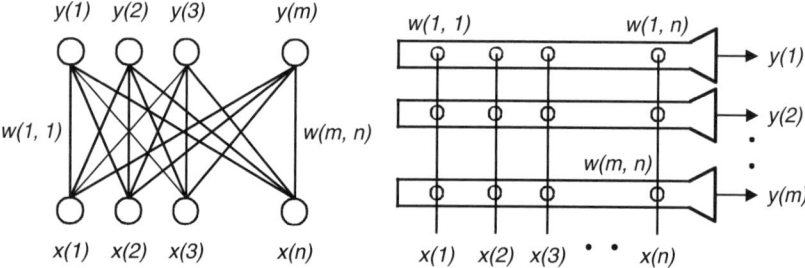

Figure 3.9 Two common depictions of the linear associator

associator only partially and in doing so have often introduced additional difficulties. However, there is an alternative route to better performance. This is the rejection of the use of the inner product in the computation of the output vector, which leads to a group of new and improved nonlinear associators.

3.2 NONLINEAR ASSOCIATORS

3.2.1 The nonlinear associative neuron group

The operation of a group of nonlinear associators is discussed here with the aid of a more general associator concept, the nonlinear associative neuron group of Figure 3.10. This associative neuron group may utilize various associative neurons, as well as the Haikonen associative neuron with certain benefits.

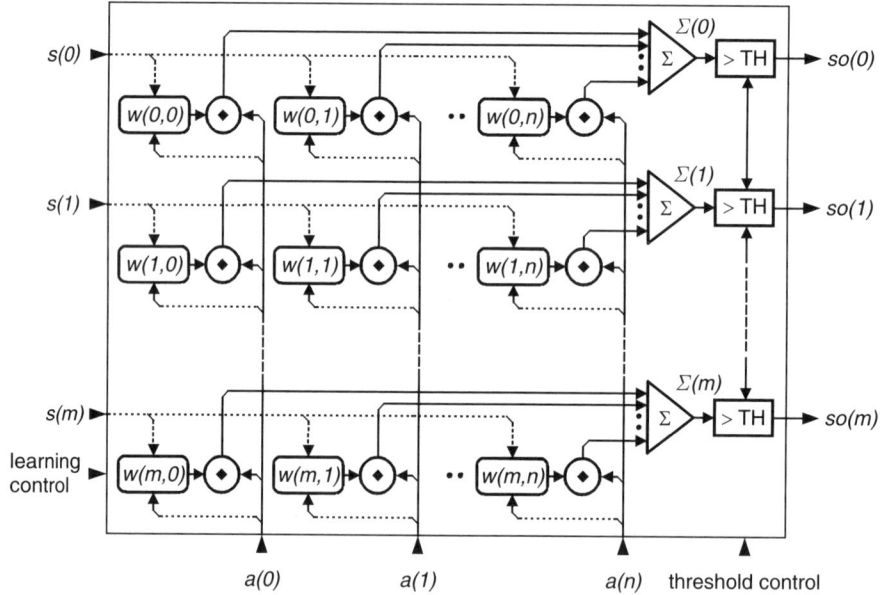

Figure 3.10 The associative neuron group

The associative neuron group of Figure 3.10 accepts the vectors $S = \{s(0), s(1), \ldots, s(m)\}$ and $A = \{a(0), a(1), \ldots, a(n)\}$ as the inputs and provides the vector $SO = \{so(0), so(1), \ldots, so(n)\}$ as the output. The weight values are depicted as $w(i, j)$ and are determined during learning by the coincidences of the corresponding $s(i)$ and $a(j)$ signals. After learning, the input vector $\{s(0), s(1), \ldots, s(m)\}$ has no further influence on the operation of the network. Learning is allowed only when the 'learning control' signal is on.

After learning, the network is able to evoke the input vector $\{s(0), s(1), \ldots, s(m)\}$ as the output with the originally associated $\{a(0), a(1), \ldots, a(n)\}$ vector or with a vector that is reasonably close to it. For the sake of clarity the evoked output vector is marked $\{so(0), so(1), \ldots, so(m)\}$.

Generally, the output of this neuron group with a given associative input vector $\{a(0), a(1), \ldots, a(n)\}$ can be computed via the computation of evocation sums $\Sigma(i)$ and comparing these sums to a set threshold TH as follows:

$$\Sigma(0) = w(0,0)\blacklozenge a(0) + w(0,1)\blacklozenge a(1) + \cdots + w(0,n)\blacklozenge a(n)$$
$$\Sigma(1) = w(1,0)\blacklozenge a(0) + w(1,1)\blacklozenge a(1) + \cdots + w(1,n)\blacklozenge a(n) \quad (3.14)$$
$$\ldots$$
$$\Sigma(m) = w(m,0)\blacklozenge a(0) + w(m,1)\blacklozenge a(1) + \cdots + w(m,n)\blacklozenge a(n)$$

or

$$\Sigma(i) = \Sigma w(i, j)\blacklozenge a(j)$$

where the summing index j runs from 1 to n.

The output $so(i)$ is determined by comparing the evocation sum $\Sigma(i)$ to the threshold value TH:

IF $\Sigma(i) < TH$ THEN $so(i) = 0$

IF $\Sigma(i) \geq TH$ THEN $so(i) = 1$

where

$\Sigma(i)$ = evocation sum

$so(i)$ = output signal ('evoked input signal')

$a(j)$ = associative input signal

$\blacklozenge$ = computational operation

$w(i, j)$ = association weight value

TH = threshold value

Traditionally, multiplication has been used as the computational operation $\blacklozenge$. In that case the evocation sum $\Sigma(i)$ is the inner product of the weight matrix and the associative input vector. However, other possibilities for the computational operation exist and will be presented in the following.

Various nonlinear associators may be realized by the nonlinear associative neuron group. Here the operation of these associators is illustrated by practical examples. In these examples the associative input vector has three bits. This gives only eight different input vectors $\{a(0), a(1), a(2)\}$ and thus the maximum number of so signals is also eight. In this limited case the complete response of an associator can be tabulated easily.

3.2.2 Simple binary associator

The simple binary associator utilizes multiplication as the computational operation $\blacklozenge$:

$$w \blacklozenge a = w^* a \qquad (3.15)$$

The output is determined by the Winner-Takes-All principle:

$$\text{IF } \Sigma(i) < max\{\Sigma(i)\} \text{ THEN } so(i) = 0$$
$$\text{IF } \Sigma(i) = max\{\Sigma(i)\} \text{ THEN } so(i) = 1$$

In this case both the weight matrix values $w(i,j)$ and the associative input vector values $a(i)$ are binary and may only have the values of zero or one. The output signals are $so(i)$: $so(0), \ldots, so(7)$. Likewise, there are only eight different associative input vectors $A = \{a(0), a(1), a(2)\}$ and are given in the bottom row. This results in a weight value matrix with eight rows and three columns. Table 3.1 gives the complete response of the corresponding associative neuron group; there are no further cases as all combinations are considered. In the table the resulting evocation sum $\Sigma(i) = w(i, 0) \blacklozenge a(0) + w(i, 1) \blacklozenge a(1) + w(i, 2) \blacklozenge a(2)$ for each A and the index i is given in the corresponding column. In practice the indexes i and j would be large. However, the conclusions from this example would still apply. This table corresponds to the associative neuron group of Figure 3.10.

An example will illustrate the contents of Table 3.1. Let $i = 3$ and $A = 001$ (see the bottom row). The third row of the weight matrix is 011. The $w \blacklozenge a$ rule is given in the left side box and in this case specifies simple multiplication. Thus the evocation sum for the third row (and the $so(3)$ signal) will be

$$\Sigma(3) = w(3, 0) \blacklozenge a(0) + w(3, 1) \blacklozenge a(1) + w(3, 2) \blacklozenge a(2) = 0^*0 + 1^*0 + 1^*1 = 1$$

According to the threshold rule each associative input vector $A = \{a(0), a(1), a(2)\}$ will evoke every signal $so(i)$ whose evocation sum $\Sigma(i)$ value exceeds the set threshold. This threshold should be set just below the maximum computed evocation sum $\Sigma(i)$ for the corresponding associative input vector A. In Table 3.1 the winning evocation sum $\Sigma(i)$ values for each A vector are circled. For

NONLINEAR ASSOCIATORS 27

Table 3.1 An example of the simple binary associator

w	a	w◆a
0	0	0
0	1	0
1	0	0
1	1	1

i	w(i,0)	w(i,1)	w(i,2)	Σ(i)	Σ(i)	Σ(i)	Σ(i)	Σ(i)	Σ(i)	Σ(i)	Σ(i)
0	0	0	0	0	0	0	0	0	0	0	0
1	0	0	1	0	**1**	0	1	0	1	0	1
2	0	1	0	0	0	**1**	1	0	0	1	1
3	0	1	1	0	**1**	**1**	**2**	0	1	1	2
4	1	0	0	0	0	0	0	**1**	1	1	1
5	1	0	1	0	**1**	0	1	**1**	**2**	1	2
6	1	1	0	0	0	**1**	1	**1**	1	**2**	2
7	1	1	1	0	**1**	**1**	**2**	**1**	**2**	**2**	**3**
A = a(0)	a(1)	a(2)		000	001	010	011	100	101	110	111

instance, the A vector 001 evokes four signals, namely $so(1)$, $so(3)$, $so(5)$ and $so(7)$ as the output with the same evocation sum $\Sigma(i)$ value: $\Sigma(1) = \Sigma(3) = \Sigma(5) = \Sigma(7) = 1$.

Generally, it can be seen that in the simple binary associator each input vector A evokes several $so(i)$ signals with equal evocation sum values, namely those ones where the A vector 'ones' match those of the corresponding row i of the weight matrix $\{w(i, 0), w(i, 1), w(i, 2)\}$. This is the mechanism that causes the evocation of unwanted responses and the apparent early filling up of the memory. The appearance of unwanted responses is also called interference.

A practical example illuminates the interference problem. Assume that two different figures are to be named associatively. These figures are described by their constituent features, component lines, as depicted in Figure 3.11.

The first figure, 'corner', consists of two perpendicular lines and the presence of these lines is indicated by setting $a(0) = 1$, $a(1) = 1$ and $a(2) = 0$. The second figure, 'triangle', consists of three lines and their presence is indicated by setting $a(0) = 1$, $a(1) = 1$ and $a(2) = 1$. A simple binary associator weight value matrix can now be set up (Table 3.2).

In Table 3.2 the $s(0)$ signal corresponds to the name 'corner' and the $s(1)$ corresponds to 'triangle'. It is desired that whenever the features of either the figure 'corner' or 'triangle' are presented the corresponding name and only that would be evoked. However, it can be seen that the features of the figure 'corner'

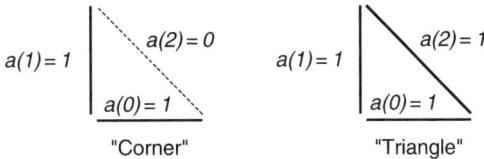

Figure 3.11 Figures and their features in the interference example

Table 3.2 An example of the interference in the simple binary associator

i	w(i,0)	w(i,1)	w(i,2)	Σ(i)
'Corner' 0	1	1	0	2
'Triangle' 1	1	1	1	2
	1	1	0	
	a(0)	a(1)	a(2)	

($a(0) = 1$, $a(1) = 1$) will lead to equal evocation sums ($\Sigma(0) = \Sigma(1) = 2$) leading to ambiguity; the simple binary associator cannot resolve these figures. This results from the fact that the features of the figure 'corner' are a subset of the features of 'triangle', and hence the name 'subset interference' (Haikonen, 1999b).

It can be seen that unwanted responses can be avoided if only mutually orthogonal rows in the weight matrix are allowed. (Any two vectors are orthogonal if their inner product is zero. In this case suitable orthogonal vectors would be the vectors {0,0,1}, {0,1,0} and {1,0,0}, which would constitute the rows in the weight matrix. Thus this associator network would only be able to resolve three suitably selected patterns.) However, the simple binary associator would be suited for applications where all signals that have 'ones' in given weight matrix positions are searched.

3.2.3 Associator with continuous weight values

The operation and capacity of the simple binary associator may be improved if continuous weight values are allowed, as indicated in Table 3.3. The modified associator is no longer a binary associator.

Table 3.3 An example of the basic associator with continuous weight values

w	a	w♦a
0	0	0
0	1	0
>0	0	0
>0	1	wa

i	w(i,0)	w(i,1)	w(i,2)	Σ(i)	Σ(i)	Σ(i)	Σ(i)	Σ(i)	Σ(i)	Σ(i)	Σ(i)
0	0	0	0	0	0	0	0	0	0	0	0
1	0	0	1	0	(1)	0	1	0	1	0	1
2	0	1	0	0	0	(1)	1	0	0	1	1
3	0	0.9	0.9	0	0.9	0.9	(1.8)	0	0	0.9	1.8
4	1	0	0	0	0	0	0	(1)	1	1	1
5	0.9	0	0.9	0	0.9	0	0.9	0.9	(1.8)	0.9	1.8
6	0.9	0.9	0	0	0	0.9	0.9	0.9	0.9	(1.8)	1.8
7	0.8	0.8	0.8	0	0.8	0.8	1.6	0.8	1.6	1.6	(2.4)
A = a(0)	a(1)	a(2)	0 0 0	0 0 1	0 1 0	0 1 1	1 0 0	1 0 1	1 1 0	1 1 1

NONLINEAR ASSOCIATORS

This associator seems to solve the subset interference problem, at least in this example, but in doing so leads to another problem, how to compute the weight values. Obviously in a more general case the weight values would have to be adjusted and tweaked against each other. This easily leads to iterative learning algorithms and training with a large number of examples. This, incidentally, would be similar to the traditional artificial neural network approach. Here, however, that kind of approach is not desired nor followed; instead other methods that conserve the binary quality of the weight values are considered.

3.2.4 Bipolar binary associator

An interesting variation of the simple binary associator can be created if instead of the zeros and ones the inputs and weights may have the values of -1 and $+1$. The computational operation ♦ and the threshold condition will be the same as those for the simple binary associator:

$$w \blacklozenge a = w^* a$$

The output is determined with the threshold value of $max\{\Sigma\}$:

$$\text{IF } \Sigma(i) < max\{\Sigma(i)\} \text{ THEN } so(i) = 0$$

$$\text{IF } \Sigma(i) \geq max\{\Sigma(i)\} \text{ THEN } so(i) = 1$$

It can be seen in Table 3.4 that the evocation sum will equal the number of signals in the associative input vector when it matches a weight matrix row. This associator executes effectively a comparison operation between each $a(i)$ and $w(i,j)$, which

Table 3.4 An example of the bipolar binary associator

w	a	w♦a
-1	-1	1
-1	1	-1
1	-1	-1
1	1	1

i	w(i,0)	w(i,1)	w(i,2)	Σ(i)	Σ(i)	Σ(i)	Σ(i)	Σ(i)	Σ(i)	Σ(i)	Σ(i)
0	-1	-1	-1	(3)	1	1	-1	1	-1	-1	-3
1	-1	-1	1	1	(3)	-1	1	-1	1	-3	-1
2	-1	1	-1	1	-1	(3)	1	-1	-3	1	-1
3	-1	1	1	-1	1	1	(3)	-3	-1	-1	1
4	1	-1	-1	1	-1	-1	-3	(3)	1	1	-1
5	1	-1	1	-1	1	-3	-1	1	(3)	-1	1
6	1	1	-1	-1	-3	1	-1	1	-1	(3)	1
7	1	1	1	-3	-1	-1	1	-1	1	1	(3)
A=a(0)	a(1)	a(2)		-1 -1 -1	-1 -1 1	-1 1 -1	-1 1 1	1 -1 -1	1 -1 1	1 1 -1	1 1 1

30 ASSOCIATIVE NEURAL NETWORKS

gives the result $+1$ whenever the $a(i)$ and $w(i,j)$ match and -1 whenever they do not match. This solves the subset interference problem. However, in practical circuit applications utilization of negative values for the synaptic weights may be a disadvantage.

3.2.5 Hamming distance binary associator

Binary associators are easy to build. Unfortunately the simple binary associator cannot give unambiguous one-to-one correspondence between the associative input vector A and the input signal $s(i)$ if the weight values of one and zero only are used. The operation of the associator would be greatly improved if a given associative input vector would evoke one and only one signal $so(i)$ without any tweaking of the weight values. The sought improvement can be realized if the inner product operation is replaced by the measurement of similarity between the associative input signal vector A and each row of the weight matrix W. A measure of the similarity of two vectors or binary strings is the Hamming distance. The Hamming distance is defined as the number of bits that differ between two binary strings. A zero Hamming distance means that the binary strings are completely similar. Associators that compute the Hamming distance are called here Hamming distance associators.

A Hamming distance binary associator may be realized by the following computational operation, which gives the Hamming distance as a negative number:

$$w \blacklozenge a = w^*(a-1) + a^*(w-1) \tag{3.16}$$

The output is determined with the fixed threshold value of zero:

$$\text{IF } \Sigma(i) < 0 \text{ THEN } so(i) = 0$$
$$\text{IF } \Sigma(i) \geq 0 \text{ THEN } so(i) = 1$$

It can be seen in Table 3.5 that the Hamming distance binary associator is a perfect associator; here each associative input vector A evokes one and only one output signal $so(i)$. Moreover, the resulting sum value $\Sigma(l)$ indicates the Hamming distance between the associative input vector A and the corresponding row in the weight matrix W. Thus, if the best match is rejected, the next best matches can easily be found. It can also be seen that the example constitutes a perfect binary three-line to eight-line converter if a fixed threshold between -1 and 0 is used. In general this Hamming distance associator operates as a binary n-line to 2^n-line converter.

3.2.6 Enhanced Hamming distance binary associator

The previously described Hamming distance associator also associates the zero A vector (0 0 0) with the output signal $so(0)$. This is not always desirable and can be avoided by using the enhanced computational operation:

NONLINEAR ASSOCIATORS 31

Table 3.5 An example of the Hamming distance binary associator

			i	$w(i,0)$	$w(i,1)$	$w(i,2)$	$\Sigma(i)$	$\Sigma(i)$	$\Sigma(i)$	$\Sigma(i)$	$\Sigma(i)$	$\Sigma(i)$	$\Sigma(i)$	$\Sigma(i)$
w	a	$w\blacklozenge a$	0	0	0	0	(0)	−1	−1	−2	−1	−2	−2	−3
0	0	0	1	0	0	1	−1	(0)	−2	−1	−2	−1	−3	−2
0	1	−1	2	0	1	0	−1	−2	(0)	−1	−2	−3	−1	−2
1	0	−1	3	0	1	1	−2	−1	−1	(0)	−3	−2	−2	−1
1	1	0	4	1	0	0	−1	−2	−2	−3	(0)	−1	−1	−2
			5	1	0	1	−2	−1	−3	−2	−1	(0)	−2	−1
			6	1	1	0	−2	−3	−1	−2	−1	−2	(0)	−1
			7	1	1	1	−3	−2	−2	−1	−2	−1	−1	(0)
			$A=a(0)$	$a(1)$	$a(2)$	0 0 0	0 0 1	0 1 0	0 1 1	1 0 0	1 0 1	1 1 0	1 1 1

$$w\blacklozenge a = w^*(a-1) + a^*(w-1) + w^* a \quad (3.17)$$

The output is determined by the Winner-Takes-All principle:

IF $\Sigma(i) < max\{\Sigma(i)\}$ THEN $so(i)=0$

IF $\Sigma(i) = max\{\Sigma(i)\}$ THEN $so(i)=1$

This enhanced Hamming distance binary associator (Table 3.6) allows the rejection of the zero–zero association by threshold control.

3.2.7 Enhanced simple binary associator

The Hamming and enhanced Hamming distance binary associators call for more complicated circuitry than the simple binary associator. Therefore the author of this book has devised another binary associator that has an almost similar performance to the Hamming and enhanced Hamming distance binary associators, but is very easy to implement in hardware. This associator utilizes the following computational operation:

$$w\blacklozenge a = w^*(a-1) + w^* a = w^*(2a-1) \quad (3.18)$$

The output is determined by the Winner-Takes-All principle:

IF $\Sigma(i) < max\{\Sigma(i)\}$ THEN $so(i)=0$

IF $\Sigma(i) = max\{\Sigma(i)\}$ THEN $so(i)=1$

The response of the enhanced simple binary associator is presented in Table 3.7.

Table 3.6 An example of the enhanced Hamming distance binary associator

	w	a	w♦a		i	w(i,0)	w(i,1)	w(i,2)	Σ(i)	Σ(i)	Σ(i)	Σ(i)	Σ(i)	Σ(i)	Σ(i)	Σ(i)
					0	0	0	0	(0)	−1	−1	−2	−1	−2	−2	−3
	0	0	0		1	0	0	1	−1	(1)	−2	0	−2	0	−3	−1
	0	1	−1		2	0	1	0	−1	−2	(1)	0	−2	−3	0	−1
	1	0	−1		3	0	1	1	−2	0	0	(2)	−3	−1	−1	1
	1	1	1		4	1	0	0	−1	−2	−2	−3	(1)	0	0	−1
					5	1	0	1	−2	0	−3	−1	0	(2)	−1	1
					6	1	1	0	−2	−3	0	−1	0	−1	(2)	1
					7	1	1	1	−3	−1	−1	1	−1	1	1	(3)
					A=a(0)	a(1)	a(2)		000	001	010	011	100	101	110	111

Table 3.7 An example of the enhanced simple binary associator

	w	a	w♦a		i	w(i,0)	w(i,1)	w(i,2)	Σ(i)	Σ(i)	Σ(i)	Σ(i)	Σ(i)	Σ(i)	Σ(i)	Σ(i)
					0	0	0	0	(0)	0	0	0	0	0	0	0
	0	0	0		1	0	0	1	−1	(1)	−1	0	−1	0	−1	1
	0	1	0		2	0	1	0	−1	−1	(1)	0	−1	−1	1	1
	1	0	−1		3	0	1	1	−2	0	0	(2)	−2	0	0	2
	1	1	1		4	1	0	0	−1	−1	−1	−1	(1)	1	1	1
					5	1	0	1	−2	0	−2	0	0	(2)	0	2
					6	1	1	0	−2	−2	0	0	0	0	(2)	2
					7	1	1	1	−3	−1	−1	1	−1	1	1	(3)
					A=a(0)	a(1)	a(2)		000	001	010	011	100	101	110	111

3.3 INTERFERENCE IN THE ASSOCIATION OF SIGNALS AND VECTORS

The previous examples relate to the association of a binary vector A with a single signal (grandmother signal) $s(i)$ and, consequently, the evocation of the corresponding single signal $so(i)$ out of many by the associated input vector A. It was seen

INTERFERENCE IN THE ASSOCIATION OF SIGNALS AND VECTORS   33

$$\begin{array}{c} so(0) \\ so(1) \\ so(2) \end{array} \begin{bmatrix} 1 & 0 & 0 \\ 0 & 1 & 0 \\ 0 & 0 & 1 \end{bmatrix}$$

$a(0)\ a(1)\ a(2)$

Figure 3.12   An example of the $(1 \to 1)$ weight matrix

$$\begin{array}{c} so(0) \\ so(1) \\ so(2) \end{array} \begin{bmatrix} 1 & 0 & 1 \\ 1 & 1 & 1 \\ 0 & 0 & 1 \end{bmatrix}$$

$a(0)\ a(1)\ a(2)$

Figure 3.13   An example of the $(1 \to n)$ weight matrix

that the simple binary associator cannot perform this operation perfectly. It was also seen that there are other associators that can do it.

In general the following association cases exist, which are considered via simple examples:

1. The association of one signal with one signal $(1 \to 1)$, an example of the weight matrix. In this case only one signal of the associative input signals $a(0)$, $a(1)$, $a(2)$ can be nonzero at a time (the accepted vectors would be $\{1, 0, 0\}$, $\{0, 1, 0\}$, $\{0, 0, 1\}$). Consequently, the inspection of the weight matrix of Figure 3.12 reveals that the associative evocation can be performed without any interference because only one output signal is evoked and no input signal may evoke false responses.

2. The association of one signal with many signals $(1 \to n)$ (a 'grandmother signal' with a vector with n components), an example of the weight matrix. Also in this case only one of the associative input signals $a(0)$, $a(1)$, $a(2)$ can be nonzero at a time. Now, however, each associative input signal $a(i)$ may evoke one or more of the output signals $s(i)$. Inspection of the weight matrix of Figure 3.13 reveals that the associative evocation can again be performed without any interference. Only the intended output signals $so(i)$ are evoked and no input signal may evoke false responses.

3. The association of many signals (a vector with n components) with one signal $(n \to 1)$, an example of the weight matrix. In this case an associative input vector evokes one and only one of the possible output signals $so(i)$. This case was discussed earlier and it was concluded that this associative operation can be performed faultlessly by any of the enhanced associators (see Figure 3.14).

4. The association of many signals with many signals (vectors with vectors) $(m \to n)$. The eight possibilities given by a binary three-bit associative input vector $a(i)$ can be depicted by a three-bit output vector $so(i)$. Thus the associator would provide a mapping between the associative input vector $a(i)$ and the output vector $so(i)$, as shown in Figure 3.15.

34 ASSOCIATIVE NEURAL NETWORKS

$$\begin{array}{c} so(1) \\ so(2) \\ so(3) \\ \vdots \\ so(7) \end{array} \begin{bmatrix} 0 & 0 & 1 \\ 0 & 1 & 0 \\ 0 & 1 & 1 \\ & & \\ 1 & 1 & 1 \end{bmatrix}$$
$$a(0)\ a(1)\ a(2)$$

Figure 3.14 An example of the $(n \rightarrow 1)$ weight matrix

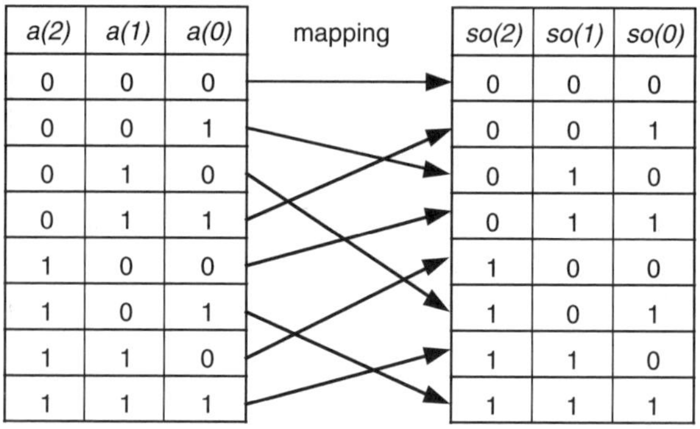

a(2)	a(1)	a(0)	mapping	so(2)	so(1)	so(0)
0	0	0		0	0	0
0	0	1		0	0	1
0	1	0		0	1	0
0	1	1		0	1	1
1	0	0		1	0	0
1	0	1		1	0	1
1	1	0		1	1	0
1	1	1		1	1	1

Figure 3.15 A mapping between the associative input vector and the output vector

Inspection of Figure 3.15 reveals that when the mapping of the 0,0,0 → 0,0,0 is fixed, seven vectors remain to be shuffled. This gives 7! (=5040) possibilities, or in a general case $(2^*n - 1)!$ different mappings if n is the number of bits in the vector. The question is: Is it possible to find a suitable weight matrix for every possible mapping? The answer is no. For instance, in the example of Figure 3.15 inspection reveals that $a(0)$, $a(1)$ and $a(2)$ are each associated with $so(0)$, $so(1)$ and $so(2)$. This would lead to a weight matrix where each individual weight would have the value of 1 and consequently every associative input vector would evoke the same output: $so(0) = so(1) = so(2) = 1$. Obviously mappings that lead to all-ones weight matrices will not work.

However, as seen before, the mappings $(m \rightarrow 1)$ and $(1 \rightarrow n)$ can be performed without interference. Therefore the mapping $(m \rightarrow n)$ may be executed in two steps, $(m \rightarrow 1)$, $(1 \rightarrow n)$, as shown in Figure 3.16.

The $(m \rightarrow 1 \rightarrow n)$ mapping will succeed if an enhanced associator is used for mapping 1. The simple binary associator is sufficient for mapping 2. The mapping structure of Figure 3.16 can be understood as an m-byte random access memory, where the byte width is 3. In this interpretation the vector $\{a(0), a(1), a(2)\}$ would be the binary address and the vector $\{so(0), so(1), so(2)\}$ would be the data. Mapping 1 would operate as an m-line to 2^m-line converter and map the binary address into the physical memory location address. Each memory location would contain the 3-bit data. However, here is the difference: a random access memory address points always to one and only one actual memory location while the associative neuron group system can find a near matching location, if all possible vectors $\{a(0), a(1), a(2)\}$ are not used. The associative neuron group has thus a built-in classification capacity.

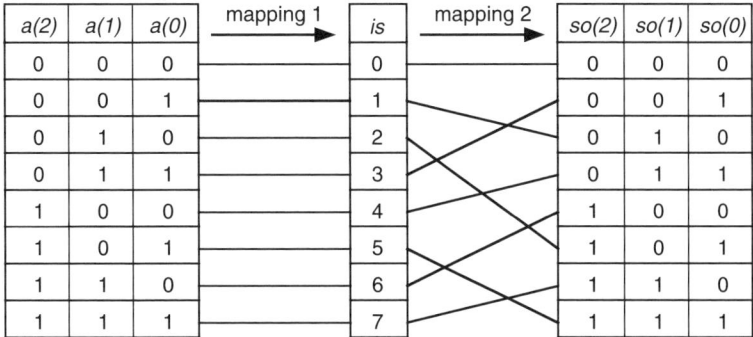

Figure 3.16 The $m \rightarrow 1 \rightarrow n$ mapping structure, with $m=n=3$

3.4 RECOGNITION AND CLASSIFICATION BY THE ASSOCIATIVE NEURON GROUP

The associative neuron group can be used to detect, recognize and classify given associative input vectors and thereby entities that these vectors are set to represent. Consider the rule for the simple binary associator:

$$\text{IF } w(i,0)^*a(0) + w(i,1)^*a(1) + \cdots + w(i,n)^*a(n) \geq TH \text{ THEN } so(i) = 1 \quad (3.19)$$

If the threshold TH equals to the number of the weight values $w(i,j) = 1$ then $so(i)$ can only have the value 1 if all the corresponding associative inputs are 1, $a(j) = 1$. The rule (3.19) now executes the logical AND operation (see also Valiant, 1994, p.113):

$$so(i) = a(0) \text{ AND } a(1) \text{ AND } \ldots \text{ AND } a(n) \quad (3.20)$$

This property may be used for entity recognition whenever the entity can be defined via its properties, for instance:

$$cherry = round \text{ AND } small \text{ AND } red$$

or

$$so = a(0) \text{ AND } a(1) \text{ AND } a(2)$$

where

$so = 1$ for *cherry*   ELSE $so = 0$
$a(0) = 1$ for *round*   ELSE $a(0) = 0$
$a(1) = 1$ for *small*   ELSE $a(1) = 0$
$a(2) = 1$ for *red*   ELSE $a(2) = 0$

36 ASSOCIATIVE NEURAL NETWORKS

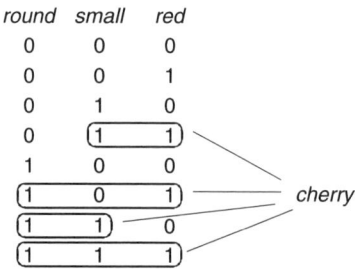

Figure 3.17 Possible property vectors for 'cherry'

This would correspond to the threshold value $TH = 3$. In this way the associative neuron group may be used to detect, recognize and name given entities.

It can be seen that all constituent properties do not have to be present if the threshold value is lowered. For instance, if in the above example the threshold is lowered to the value 2 then only two properties suffice for the recognition. *Cherry* will be detected if one of the following imperfect conditions is present: *round* AND *red* or *round* AND *small* or *small* AND *red*. These constitute here a 'close enough' condition (Figure 3.17). This imperfect or soft AND operation can also be seen as classification; here four somewhat similar vectors are taken as examples of the class *cherry*.

This kind of associative classification is very useful. Already one representative example may be enough for the learning of a class (in this example the vector 111). Thereafter all examples that are close enough are taken to belong to that class.

However, it should also be possible to reclassify any examples as new information comes in. Here, for instance, the combination of properties {*round* and *small*}, the vector 110, might be taken to represent a *marble*. Would the neuron group now be able to resolve the classes of *cherry* and *marble* correctly? Simple inspection shows that the linear binary associator cannot do this. In Table 3.8 the evocation sums in the binary linear associator for all combinations of the properties *round, small* and *red* {$a(0), a(1), a(2)$} are tabulated. It can be seen that the combination {*round* and *small*}, the vector 110 for *marble*, gives the same evocation sum 2 for $so(0) = marble$ and $so(1) = cherry$, and thus the neuron group cannot resolve between these no matter which threshold strategy is used. The reason for this failure is obvious; the ones in the vector 110 (*marble*) are a subset of the ones in the vector 111 (*cherry*) and the $w \blacklozenge a = w*a$ operation of the binary linear associator is not able to detect that the vector 110 is a full match for the class *marble* and only a partial match for the class *cherry*. This failure mode is called here 'subset interference'.

The subset interference in the simple binary associator can be avoided or diminished by the following methods: (a) by allowing only mutually orthogonal rows in the weight matrix, (b) by using additional property signals $a(i)$ (for instance those that relate to the number of ones in the property vector A), (c) by sparse coding, using very long A and W vectors where the number of ones is small compared to the number of zeros.

The subset interference can be avoided by using associators with a more complicated $w \blacklozenge a$ operation. Table 3.9 gives the evocation sums in the enhanced Hamming

RECOGNITION AND CLASSIFICATION   37

Table 3.8 Subset interference of the simple binary associator

w	a	w♦a		i	w(i,0)	w(i,1)	w(i,2)	Σ(i)	Σ(i)	Σ(i)	Σ(i)	Σ(i)	Σ(i)	Σ(i)	
0	0	0		0	1	1	0	0	1	1	1	1	(2)	2	Marble
0	1	0													
1	0	0		1	1	1	1	1	1	(2)	1	(2)	(2)	3	Cherry
1	1	1													
				A=a(0)	a(1)	a(2)	001	010	011	100	101	110	111		
				Round	Small	Red									

Table 3.9 The subset interference resolved using the enhanced Hamming distance associator

w	a	w♦a		i	w(i,0)	w(i,1)	w(i,2)	Σ(i)	Σ(i)	Σ(i)	Σ(i)	Σ(i)	Σ(i)	Σ(i)	
0	0	0		0	1	1	0	−3	0	−1	0	−1	(2)	1	Marble
0	1	−1													
1	0	−1		1	1	1	1	−1	−1	(1)	−1	(1)	1	3	Cherry
1	1	1													
				A=a(0)	a(1)	a(2)	001	010	011	100	101	110	111		
				Round	Small	Red									

distance associator for all combinations of the properties *round, small* and *red* {a(0), a(1), a(2)}. It can be seen that by using the Winner-Takes-All threshold strategy the classes of *cherry* and *marble* can be properly resolved. Here the minimum threshold value 1 is used; the vectors 011, 101 and 111 become correctly to represent *cherry* and the vector 110 represents also correctly *marble*. If the minimum threshold were to be lowered to 0 then the vectors 010 and 100 would also come to represent *marble*, again correctly. The reason for the improved operation is that the threshold rule (3.19) no longer reduces into the simple AND operation; instead it also considers the missing properties.

Unfortunately the subset interference is not the only interference mechanism in associators. Another interference mechanism is the so-called exclusive-OR (EXOR) problem. This problem arises when for a given class there are two properties that can appear alone but not together – this or that but not both. If these properties are marked as $a(0)$ and $a(1)$ then the condition for the class would be the logical exclusive-Or operation between the two properties, $a(0)$ EXOR $a(1)$. Suppose that an enhanced Hamming distance associator is set up to resolve the two classes $so(0) = a(0)$ EXOR $a(1)$ and $so(1) = a(0)$ AND $a(1)$. This leads to a 2×2 weight matrix full of ones (Table 3.10).

It can be seen that the enhanced Hamming distance associator cannot resolve the two classes as the weight values for $so(0)$ and $so(1)$ are the same. Therefore additional information is required to solve the ambiguity. This information can be provided by the inclusion of an additional signal $a(2)$. This signal is set to be 1 if $a(0) + a(1) = 2$. Thus $a(2) = 1$ for $a(0)$ AND $a(1)$ and $a(2) = 0$ for $a(0)$ EXOR

ASSOCIATIVE NEURAL NETWORKS

Table 3.10 The exclusive-OR (EXOR) interference

w	a	w♦a	i	w(i,0)	w(i,1)	Σ(i)	Σ(i)	Σ(i)	
0	0	0	0	1	1	0	0	2	a(0) EXOR a(1)
0	1	−1	1	1	1	0	0	2	a(0) AND a(1)
1	0	−1		A=a(0)	a(1)	0 1	1 0	1 1	
1	1	1							

Table 3.11 The exclusive-OR (EXOR) interference solved

w	a	w♦a	i	w(i,0)	w(i,1)	w(i,2)	Σ(i)	Σ(i)	Σ(i)	
0	0	0	0	1	1	0	(0)	(0)	1	a(0) EXOR a(1)
0	1	−1	1	1	1	1	−1	−1	(3)	a(0) AND a(1)
1	0	−1		A=a(0)	a(1)	a(2)	0 1 0	1 0 0	1 1 1	
1	1	1								

$a(1)$. Note that the A vector $\{a(0), a(1), a(2)\} = \{1, 1, 0\}$ is not allowed now. This leads to the associator of Table 3.11.

Thus the exclusive-OR interference can be avoided by the introduction of additional information about the coexistence of associative input signals.

3.5 LEARNING

The determination of the weight values is called learning at the neural network level (cognitive learning is discussed later on). In the associators learning is usually based on a local calculation; each weight depends only on the presence of the signals to be associated with each other. Nonlocal calculation is frequently used in other artificial neural networks. There each weight depends also on the values of other weights, thus iterative algorithms that effectively tweak weights against each other are needed. Here only local learning is considered.

Two basic cases of associative learning are discussed here, namely instant Hebbian learning and correlative Hebbian learning. The latter enables better discrimination of entities to be associated.

3.5.1 Instant Hebbian learning

Instant Hebbian learning is fast. It associates two signals with each other instantly, with only one coincidence of these signals.

The association weight value $w(i, j)$ is computed as follows at the moment of association:

$$w(i, j) = s(i) * a(j)$$
$$\text{IF } s(i) * a(j) = 1 \text{ THEN } w(i, j) \Rightarrow 1 \text{ permanently}$$

where

$s(i)$ = input of the associative matrix; zero or one
$a(j)$ = associative input of the associative matrix; zero or one

The association weight value $w(i, j) = 1$ gained at any moment of association will remain permanent.

Instant learning is susceptible to noise. Random coincidences of signals at the moment of association will lead to false associations.

3.5.2 Correlative Hebbian learning

Instant learning can connect two entities together, for instance an object and its name. However, if the entity to be named is a subpart or a property of a larger set then instant learning will associate the name with all signals that are present at that moment, thus leading to a large number of undesired associations. Correlative learning will remedy this, but with the cost of repeated instants of association; learning takes more time.

In correlative Hebbian learning association between the intended signals is created by using several different examples, such as sets of signals that contain the desired signals as a subset. Preferably this subset should be the only common subset in these examples.

Correlative Hebbian learning is described here by using a single associative neuron with the input signal $s(t)$, the output signal $so(t)$ and the weight vector $w(i)$. Each weight value $w(i)$ is determined as a result of averaged correlation over several training examples. For this purpose a preliminary correlation sum $c(i, t)$ is first accumulated and the final weight value $w(i)$ is determined on the basis of the following sum (see rule (3.8)):

$$c(i, t) = c(i, t - 1) + 1.5 * s(t) * a(i, t) - 0.5 * s(t)$$
$$\text{IF } c(i, t) > threshold \text{ THEN } w(i) \Rightarrow 1$$

where

$c(i, t)$ = correlation sum at the moment t
$s(t)$ = input of the associative neuron at the moment t; zero or one
$a(i, t)$ = associative input of the associative neuron at the moment t; zero or one

40 ASSOCIATIVE NEURAL NETWORKS

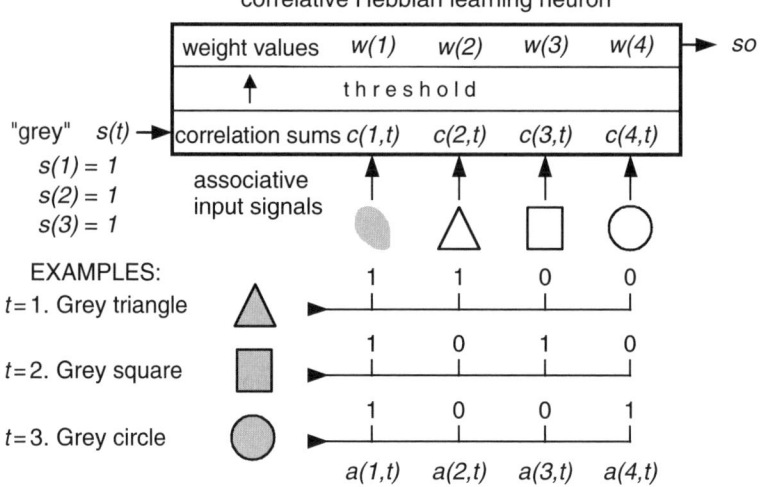

Figure 3.18 An example of correlative learning

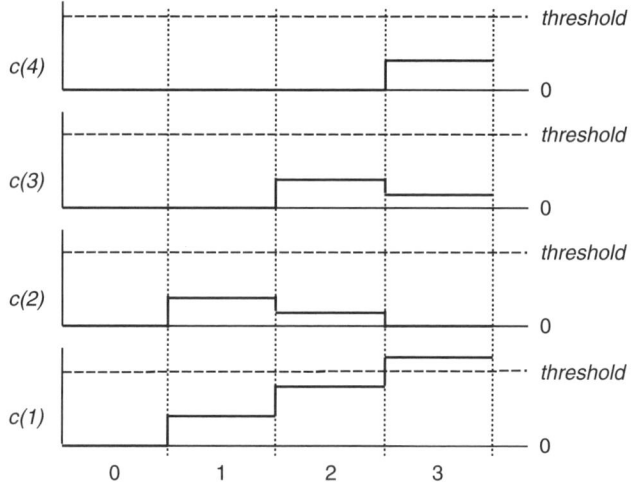

Figure 3.19 Correlation sums of the example of correlative learning

The association weight value $w(i) = 1$ gained at any moment of association will remain permanent. This rule is the same as given for the Haikonen associative neuron; other similar rules can be devised.

An example illustrates the principle of correlative learning (Figure 3.18). In Figure 3.18 the name 'grey' (the input s) is associated with the respective property <grey> with the aid of three example objects: a grey triangle, a grey square and a grey circle. The correlation sums after each training examples are depicted in Figure 3.19.

After the first training example the property <grey> and the shape <triangle> begin to be associated with 'grey'; the corresponding correlation sums $c(1, t)$ and

$c(2, t)$ rise from zero. The second training example raises further the correlation sum $c(1, t)$, the correlation between 'grey' and <grey>. At the same time the correlation sum $c(2, t)$ decreases; the system begins to 'forget' an irrelevant association. The correlation sum $c(3, t)$ rises now a little due to the coincidence between 'grey' and the shape <square>. After the third training example the correlation sum $c(1, t)$ exceeds the threshold and the weight value $w(1)$ gains the value of 1 while the other weight values $w(2)$, $w(3)$ and $w(4)$ remain at zero. The common property <grey> is now associated with the name 'grey' while no undesired associations with any of the pattern features take place. Here learning also involves forgetting.

3.6 MATCH, MISMATCH AND NOVELTY

By definition the meaning of the evoked output vector SO of the associative neuron group of Figure 3.10 is the same as that of the input vector S. In perceptive associative systems the input vector S may represent a percept while the evoked output vector SO may represent a predicted or expected percept of a similar kind. In that case the system should have means to determine how the prediction or expectation corresponds to the actual perception. This requirement leads to the match/mismatch/novelty comparison operation between the instantaneous input vector S and the output vector SO, as depicted in Figure 3.20. This comparison assumes that the Haikonen neurons are used and the switch SW is open so that the input vector S does not appear at the output (see Figure 3.3).

In Figure 3.20 the *match* condition occurs when the instantaneous input vector S is the same or almost the same as the output vector SO. The *mismatch* condition occurs when the two vectors do not match. The *novelty* condition occurs when there is no evoked output vector; the system does not have any prediction or expectation for the input vector, yet an input vector appears. Three mutually exclusive signals may thus be derived corresponding to the match, mismatch and novelty conditions. These signals may be binary, having only the values of one or zero, but graded signals may also be used. In the following, binary match, mismatch and novelty signals are assumed.

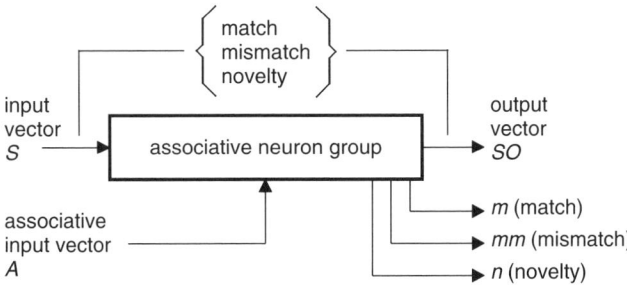

Figure 3.20 Match, mismatch and novelty as the relationship between the input vector S and the evoked output vector SO

Table 3.12 The exclusive-OR operation EXOR, $c = a$ EXOR b

a	b	c
0	0	0
0	1	1
1	0	1
1	1	0

The match/mismatch/novelty conditions may be determined by the Hamming distance between the input vector S and the output vector SO. The Hamming distance between two vectors may be computed as the sum of nonmatching bits. Here the logical exclusive-OR operation may be used to detect the nonmatching bits. The truth table for the exclusive-OR operation is given in Table 3.12.

The exclusive-OR operation generates a 'one' when its inputs do not match and 'zero' when there is a match. Thus the Hamming distance Hd between S and SO vectors can be determined as follows:

$$Hd = \Sigma(s(i) \text{ EXOR } so(i)) \qquad (3.21)$$

The match/mismatch/novelty conditions may be determined by the following rules:

$Hd \leq threshold \Rightarrow$ match condition: $m = 1$, $mm = 0$, $n = 0$

$S \neq SO$; $Hd > threshold$ AND $SO \neq 0 \Rightarrow$ mismatch condition: $m = 0$, $mm = 1$, $n = 0$

$S \neq 0$ AND $SO = 0 \Rightarrow$ novelty condition: $m = 0$, $mm = 0$, $n = 1$ \qquad (3.22)

The match/mismatch/novelty detection has many important applications, as will be seen in the following chapters. One application relates to learning control and specially to overwriting protection. Unwanted overwriting may take place no matter how large a memory or a neuron group capacity is unless some protection is provided. In this case associative evocation should be tried before learning. If the mismatch condition occurs then something else has already been learned and new associations might cause problems. Therefore learning should only be allowed during the novelty condition.

3.7 THE ASSOCIATIVE NEURON GROUP AND NONCOMPUTABLE FUNCTIONS

The fully trained associative neuron group can be compared to a device that computes the function $SO = f(A)$, where SO is the output vector and A is the input vector. In mathematics a computable function is an input-to-output mapping, which can be specified by an algorithm that allows the computation of the output when the input is given (for example, see Churchland and Sejnowski, 1992, p. 62). On the other hand, a limited size mapping can be presented as a list of input–output pairs. This kind of list can be implemented as a look-up table. This table represents the maximum number of stored bits that may be needed for the determination of the output for each possible input. A program that executes the corresponding algorithm should require less stored bits than the look-up table; otherwise it would not be computationally effective. An efficient algorithm may also be considered as a lossless compression of the original data. Random data contains the maximum information and cannot be compressed, that is represented by fewer bits than it has. Therefore, for random data no compressing computational rule can be devised and no efficient algorithms can exist.

Thus, an algorithm can be used to compute the function $SO = f(A)$ if a computational rule exists. If the output vectors SO and the input vectors A form random pairings then no rule exists and the function cannot be computed. However, in this case the input–output pairs can also be represented as a look-up table. A random access memory may store random pairings. The associative neuron group is not different if it is configured so that no interference exists, as discussed before. Thus the associative neuron group is able to execute noncomputable functions. In a perceptually learning cognitive machine the external world may translate into random pairings of vectors and this correspondence may thus be noncomputable. Consequently, rule-based modelling will be inaccurate or fail in essential points. The associative neuron group is a kind of self-learning look-up table that will adapt to the environment when no rules exist. In this context it should be noted that the cognitive system that is outlined in the following is not a look-up table. It is not a state machine either, but a system that utilizes the associative neuron groups in various ways with some additional circuits and processes, so that 'intermediate results' can be accessed and reused.

4

Circuit assemblies

4.1 THE ASSOCIATIVE NEURON GROUP

A general associative neuron group circuit is used as the basic building block for the systems and circuits that are described in the following chapters. This circuit utilizes an enhanced associator that is realized as a group of the Haikonen associative neurons. Figure 4.1 depicts the associative neuron group and its simplified drawing symbol.

The input signals for the associative neuron group are: the main signal vector $S = \{s(0), s(1), \ldots, s(m)\}$, the associative input signal vector $A = \{a(0), a(1), \ldots, a(n)\}$, the learning control signal, the threshold control signal for the associative input signal thresholds THa, the threshold control signal for the evocation output thresholds TH and the SW switch control signal. The output signals are: the output signal vector $SO = \{so(0), so(1), \ldots, so(m)\}$, the match signal m, the mismatch signal mm and the novelty signal n. The main signal inputs for $s(i)$ have their own input threshold circuits THs. Depending on the application, these thresholds may utilize linear or limiting threshold functions (see Section 3.1.4). Likewise, the associative signal inputs for $a(j)$ have also their own input threshold circuits THa. These thresholds operate according to the Winner-Takes-All principle; all associative input signals that share the maximum value are accepted. Thus the input signals may have continuous values, but the signals that are passed to the associative matrix have the discrete values of one or zero. The switch SW allows the main signal to pass through the circuit when closed. This feature is useful in some applications.

In the simplified neuron group symbol the horizontal signal lines depict S vector inputs and SO vector outputs. The vertical signal lines depict associative input vectors. For practical reasons the simplified drawing symbol for the neuron group is used in the following system diagrams. All input, output and control lines are not always necessarily drawn, but their existence should be understood.

46 CIRCUIT ASSEMBLIES

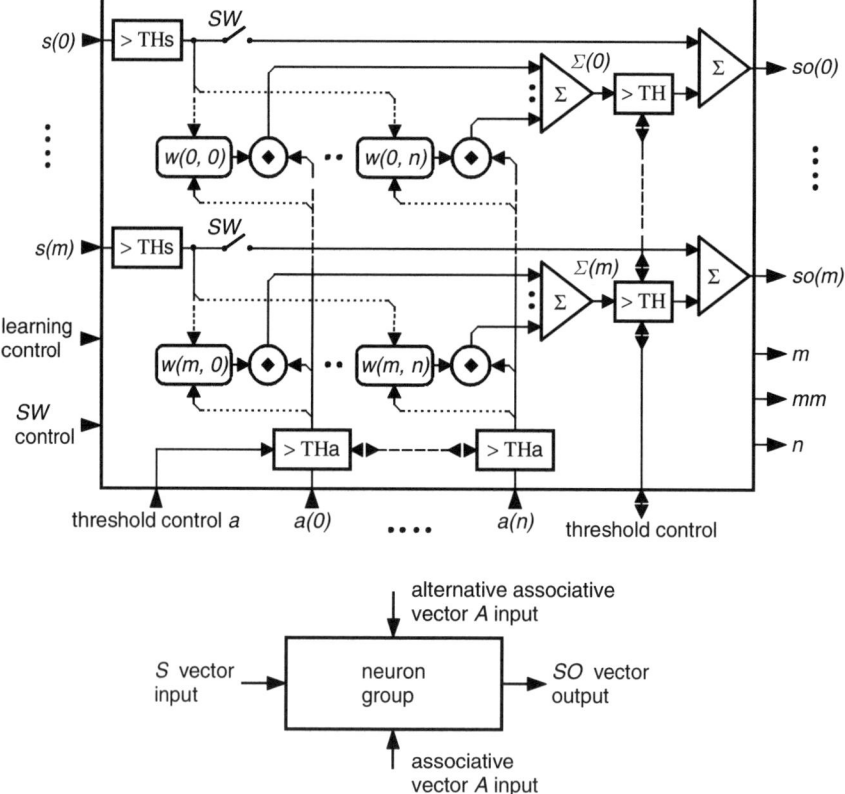

Figure 4.1 The associative neuron group and its simplified drawing symbol

4.2 THE INHIBIT NEURON GROUP

In certain applications an associative inhibition operation is needed. In that case the associative vector A inhibits the associated input vector S. In the absence of the associative input vector A all S vectors are forwarded to the output, $S = SO$. The inhibit neuron group and its simplified drawing symbol is depicted in Figure 4.2.

The inhibit neuron group is similar to the associative neuron group except for the S vector inhibit logic. The inhibit neuron group of Figure 4.2 utilizes binary logic inhibit operation. It is obvious that instead of binary logic analog switching could also be used.

4.3 VOLTAGE-TO-SINGLE SIGNAL (V/SS) CONVERSION

The associative neuron group is suited for processing on/off or 1/0 signals only. Therefore continuous signal values must be converted into 1/0 valued signals in order to make them compatible with the associative neuron group. This can be done by dividing the intensity range of the continuous signal into equal fractions and assigning a single signal to each fraction. This process is illustrated in Figure 4.3 by the help of a voltage ramp signal.

VOLTAGE-TO-SINGLE SIGNAL CONVERSION   47

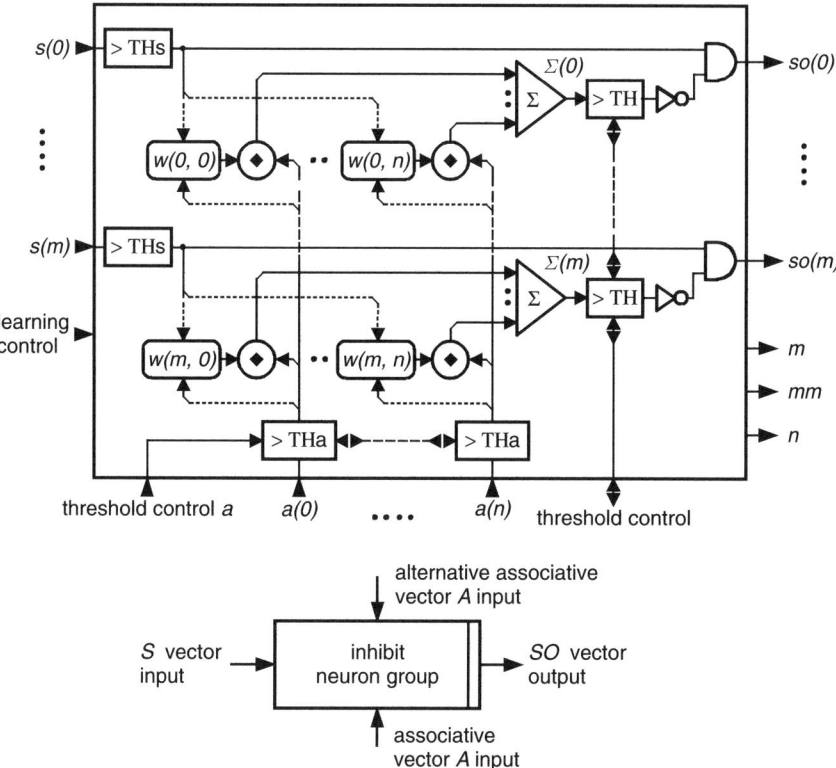

Figure 4.2 The inhibit neuron group and its simplified drawing symbol

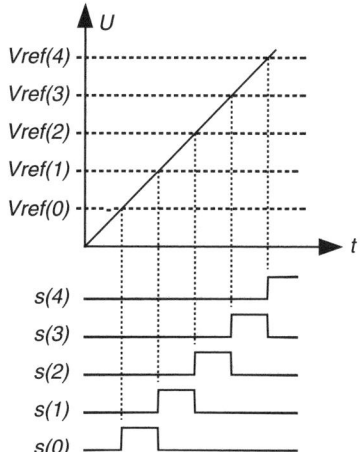

Figure 4.3 Voltage-to-single signal conversion

In Figure 4.3 the intensity range of the continuous signal is divided into equal fractions or steps with the help of evenly paced threshold voltage levels $Vref(i)$. Thus, whenever the signal value is above the threshold value $Vref(i)$ but lower than

48 CIRCUIT ASSEMBLIES

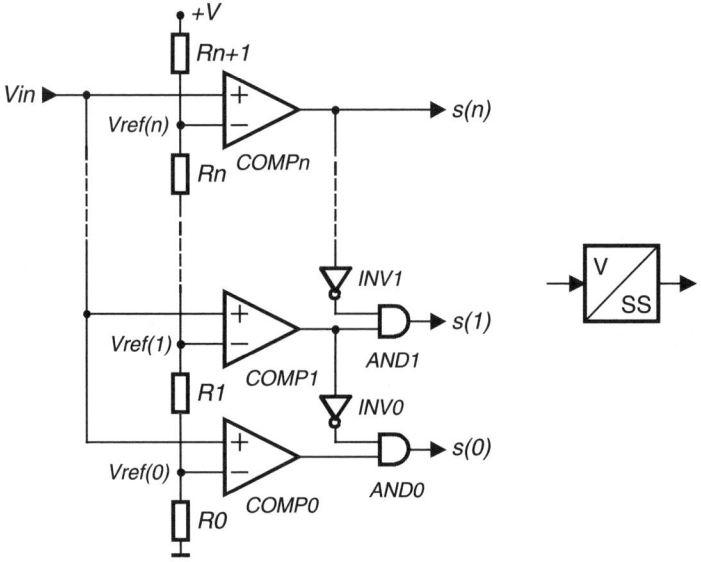

Figure 4.4 A voltage-to-single signal (V/SS) converter and its symbol

$Vref(i+1)$ the dedicated single signal $s(i)$ has the value 1 and all the other signals $s(j \neq i)$ have the value 0.

This process introduces a quantization error similar to that of an analog-to-digital conversion. Therefore the number of s signals should be chosen high enough to keep the quantization error small enough for each application. A circuit that executes the above conversion is shown in Figure 4.4.

In the circuit of Figure 4.4 the threshold voltage levels $Vref(i)$ are formed by the resistor chain $R0, \ldots, Rn+1$. The continuous input signal Vin is compared against these thresholds by the comparators $COMP0, \ldots, COMPn$. The output of each comparator is zero if the continuous input signal voltage Vin is lower than the respective threshold value and one if the voltage Vin is higher than the respective threshold value. Now, however, only one of the output signals $s(i)$ may have the value of one at any moment, namely the one that corresponds to the highest threshold value that is being exceeded by the input voltage Vin. This is secured by the *AND* gating circuits at the outputs.

Any possible significance information must be modulated on the $s(i)$ signals by other circuitry.

4.4 SINGLE SIGNAL-TO-VOLTAGE (SS/V) CONVERSION

In some applications single signals must be converted back into continuous signal values. In Figure 4.5 a simple conversion circuit is shown.

In the circuit of Figure 4.5 the input signals are $s(0) \ldots s(n)$. Only one of these signals may be non zero and positive at any given time. A non zero positive $s(i)$ signal closes the corresponding switch SWi so that the voltage Vs is coupled to the

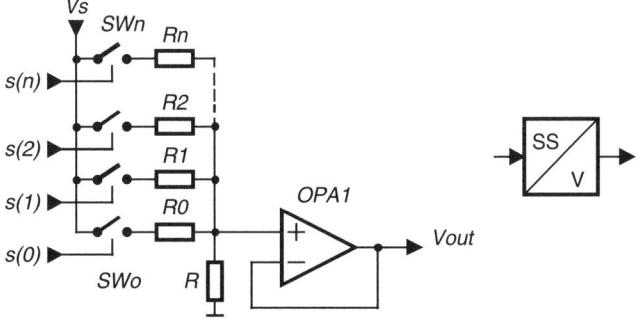

Figure 4.5 A single signal-to-voltage converter and its symbol

corresponding resistor Ri. The resulting output voltage $Vout$ will be determined by the voltage division by the resistors Ri and R:

$$Vout = Vs^*R/(Ri+R) \qquad (4.1)$$

where

$Vout =$ output voltage
$Vs =$ signal voltage corresponding to logical one

The value of each Ri must be chosen so that for each $s(i) = 1$

$$Vout = (i+1)^*\Delta V \qquad (4.2)$$

where

$\Delta V =$ step size

From Equations (4.1) and (4.2) it follows that

$$Ri = Vs^*R/((i+1)^*\Delta V) - R \qquad (4.3)$$

For example, if $\Delta V = Vs/100$ and $i = 0$ (the first step) then

$$R0 = Vs^*R/(0.01^*Vs) - R = 99R$$

4.5 THE 'WINNER-TAKES-ALL' (WTA) CIRCUIT

In certain cases the strongest signal of many parallel signals must be selected. This can be done by the so-called 'Winner-Takes-All' (WTA) circuit, which selects

50 CIRCUIT ASSEMBLIES

the strongest signal and inhibits the others. There are many possibilities for the realization of a WTA circuit, here one such circuit is described (Haikonen, 1999b). This circuit has an important benefit. Instead of a large number of inhibit lines between the individual threshold circuits for each signal, this circuit utilizes only one, the threshold control line.

Figure 4.6 depicts a Winner-Takes-All (WTA) assembly with threshold circuits for n parallel signals. The input signals are $s(0), s(1), \ldots, s(n)$ and they may have continuous voltage values between zero and some limited positive value. The corresponding output signals are $so(0), so(1), \ldots, so(n)$. The WTA assembly should pass the highest valued $s(i)$ as the output $so(i)$ and keep the other outputs at a low level. This is an analog circuit where the voltage of the winning $s(i)$ signal is passed to the output. In practical applications some supporting circuitry and modifications may be required.

The operation of the WTA circuit assembly is explained by using the uppermost circuit as the reference. The comparator $COMP0$ compares the voltage of the signal $s(0)$ to the voltage TH of the threshold control line. If the voltage of the signal $s(0)$ is higher than the highest voltage of the other signals $s(1), \ldots, s(n)$ then $s(0) > TH$. At that moment the comparator $COMP0$ acts as an analog voltage follower and forces the threshold control line voltage TH to that of the signal $s(0)$. Subsequently, all the diodes $D1, \ldots, Dn$ and Db will be reverse biased and will cut off all the other

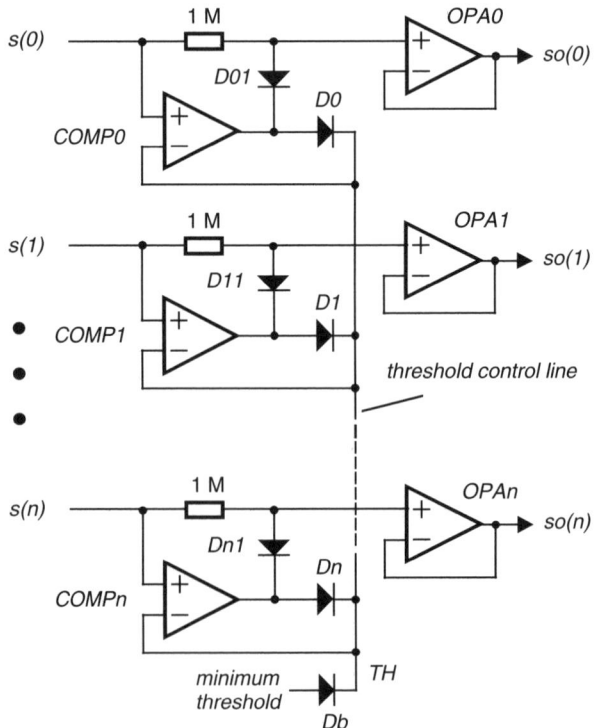

Figure 4.6 A 'Winner-Takes-All' (WTA) threshold circuit assembly

THE 'ACCEPT-AND-HOLD' CIRCUIT 51

influences on the threshold control line. The diode $D01$ will now stop conducting and the voltage at the noninverting input of the voltage follower $OPA1$ will have the value of the input voltage $s(0)$ and consequently the corresponding output $so(0)$ will equal this value $s(0)$.

If the voltage of the $s(0)$ signal is lower than the highest voltage of the other signals then $s(0) < TH$. In that case the comparator $COMP0$ output will be at the low level and the diode $D0$ will be reverse biased. The diode $D01$ will now conduct and short the noninverting input of $OPA0$. Consequently the output $so(0)$ will remain at the low level.

In this way the signal $s(i)$ with the highest voltage is selected. If there are several signals with the same high value, then they will all be selected. A minimum acceptance threshold may be set via the diode Db.

4.6 THE 'ACCEPT-AND-HOLD' (AH) CIRCUIT

The 'Accept-and-Hold' (AH) circuit is an associative circuit that accepts and holds input vectors that it has learned earlier. The circuit can be used to sort out and capture given vectors out of a stream of vectors or a temporal vector sequence: 'it takes its own vectors'. The AH circuit and its simplified depiction is presented in Figure 4.7.

The 'Accept-and-Hold' (AH) circuit of Figure 4.7 consists of two cross-connected neuron groups, which are the neuron group S and the neuron group G. The circuit receives two input vectors, S and G. The S vector is the captured one and will be output as the So vector. The SW switch (see Figure 4.1) of the neuron group S is open, so that the input vector S will not emerge as the output vector So, which can only be associatively evoked by the output vector of the neuron group G, the Go vector. In the neuron group G the SW switch is closed.

During initial learning the S and G vectors are associated with each other at the neuron group G. The G vector emerges as the output vector Go, allowing its association with the vector S at the neuron group S. Immediately thereafter the output vector So will emerge due to the associative evocation by the Go vector, and will be associated with the G vector. Now the circuit is ready to accept and hold the learned vector S. When the learned vector S (or some vector that is similar enough) enters the circuit it evokes the vector Go at the output of the neuron group G; this

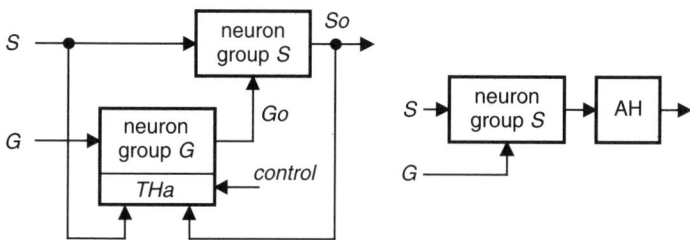

Figure 4.7 The 'Accept-and-Hold' (AH) circuit and its simplified depiction

52 CIRCUIT ASSEMBLIES

in turn evokes the output vector $So = S$ at the output of the neuron group S, which in turn again evokes the vector Go. The evocation loop $So \rightarrow Go \rightarrow So$ will sustain the output So indefinitely. This is a stabile condition, which can be reset by raising the associative input threshold level Tha at the neuron group G.

As soon as an input vector S has been captured, no new inputs must be allowed until the circuitry has been reset. This can be accomplished by the associative input threshold Tha, which is to be configured so that the active vector So inhibits any S input signals.

In this circuit the acceptance of the S vectors is connected to the G vector, which in a way grounds the general meaning of the accepted S vectors; only S vectors that are associated with G vectors are accepted. In this circuit the accepted vector So can also be evoked by the G vector. The circuit can be modified so that only the S input may evoke accepted vectors. There are also other possibilities for the realization of the AH circuit.

4.7 SYNAPTIC PARTITIONING

Sometimes several associative input vectors $A0, A1, \ldots, An$ are connected to one associative neuron group. In those cases it is useful to partition the synapses into groups that match the associative input vectors, as shown in Figure 4.8.

The neuron group with the partitioned synapse groups of Figure 4.8 executes the following evocation rule:

$$\text{IF } \Sigma w(i,j) \blacklozenge a(1,j) + \Sigma w(i, j+p) \blacklozenge a(2,j)$$
$$+ \cdots + \Sigma w(i, j+(n-1)^*p) \blacklozenge a(n,j) \geq TH$$
$$\text{THEN } so(i) = 1 \quad (4.4)$$

where

$n =$ number of the associative input groups
$p =$ number of associative inputs in any $A1, A2, \ldots, An$ group
$i =$ index for the $s(i)$ and $so(i)$ signals
$j =$ summing index; runs from 1 to p

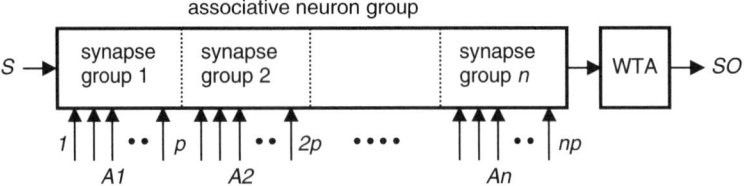

Figure 4.8 A neuron group with partitioned synapse groups

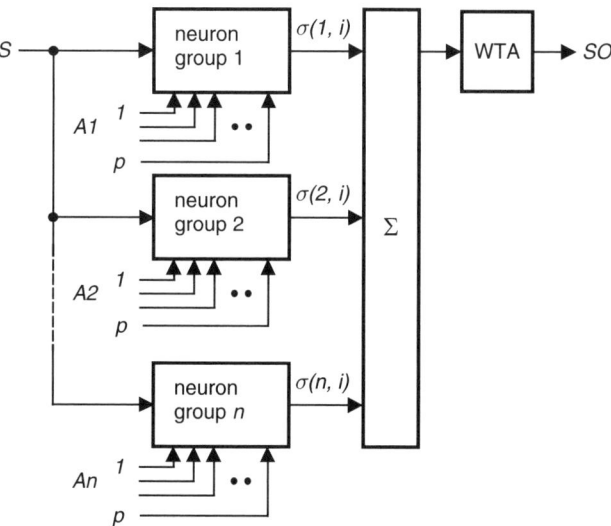

Figure 4.9 Parallel neuron groups used instead of synaptic partitioning

Here the total number of associative input signals a is $n*p$. Thus, for the associative neuron group the associative input index runs from 1 to $n*p$. However, for each associative input vector the index runs from 1 to p. In rule (4.4) the index j runs from 1 to p and therefore conforms to the associative input vector signal number. The index term $j+(n-1)*p$ makes the summing index suitable for the associative neuron group. For the A1 group the summing index for the associative neuron group runs from 1 to $j+(1-1)*p=1$ to j (1 to p). For the A2 group the summing index runs from p to $j+(2-1)*p=p$ to $j+p$ (p to $2*p$) and so on.

This function may also be executed by n parallel neuron groups, as depicted in Figure 4.9. In Figure 4.9 each neuron group accepts the signal vector S as the main input. The output threshold circuit is omitted from the individual neuron groups and the outputs consist of the raw evocation sums. These are summed together and the sums are forwarded to the WTA threshold circuit. The evocation sums and the evocation rule for the output signals are computed as follows:

$$\text{IF } \Sigma \sigma(k, i) \geq TH \text{ THEN } so(i) = 1 \qquad (4.5)$$

where

$\sigma(k, i) = \Sigma w(k, i, j) \blacklozenge a(k, j) =$ evocation sum for the signal $so(i)$ at the neuron group k
$k =$ index for the neuron group; runs from 1 to n

Thus rule (4.5) can be rewritten as:

$$\text{IF } \Sigma w(1, i, j) \blacklozenge a(1, j) + \Sigma w(2, i, j) \blacklozenge a(2, j) + \cdots + \Sigma w(n, i, j) \blacklozenge a(n, j) \geq TH$$
$$\text{THEN } so(i) = 1 \qquad (4.6)$$

54   CIRCUIT ASSEMBLIES

where

$n =$ number of neuron groups

It can be seen that the evocation rules (4.4) and (4.6) are equivalent and the circuits of Figures 4.8 and 4.9 execute the same function.

4.8 SERIAL-TO-PARALLEL TRANSFORMATION

In many cases entities are represented by temporal series of parallel distributed signal representations, signal vectors, like the phonemes of a word, words of a sentence, serially tracked subcomponents of a visual object, etc. However, for processing purposes these serially occurring representations may be needed simultaneously, at least for a short while. This can be achieved with serial-to-parallel transformation circuits. The serial-to-parallel transformation makes a number of temporally serial representations available simultaneously for a certain period of time.

In principle the serial-to-parallel transformation can be performed with two different methods, which are called here the serial method and the parallel method. If binary signal vectors are used then these methods may be realized by chains of conventional digital registers.

The shift register chain for the serial-to-parallel transformation by the serial method is presented in Figure 4.10. In Figure 4.10 each register consists of a number of parallel D flip-flops and is able to capture and store a signal vector at the moment of the timing command. The captured signal vector is then available as the output of the register until the next timing command. At each register the input vector is available as the output only after a small delay; therefore at the moment of the timing command the next register in the chain sees the previously stored output vector of

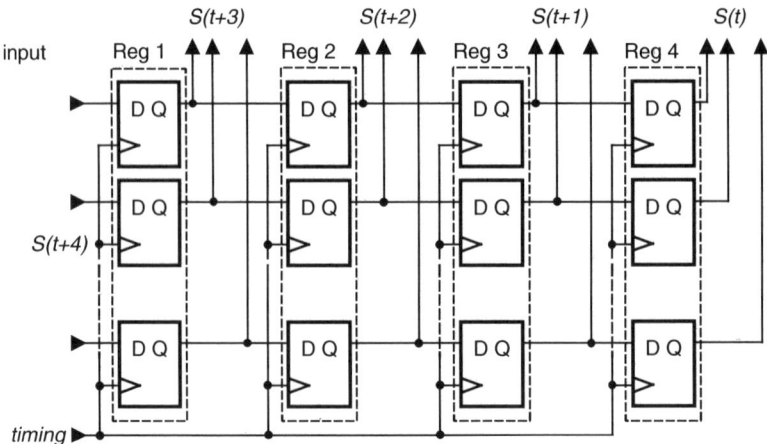

Figure 4.10   Shift register chain for the serial-to-parallel transformation by the serial method

the preceding register. Thus the register chain will be able to shift an input vector step by step through the chain and the content in the registers will flow as depicted in Figure 4.11. It can be seen that the shift register chain of Figure 4.10 with n registers will be able to present n last vectors of a temporal sequence simultaneously.

The shift register chain for the serial-to-parallel transformation by the parallel method is presented in Figure 4.12. In the parallel method a number of successive vectors $S(t)$, $S(t+1)$, $S(t+2)$, ..., etc., are captured into a number of parallel registers. The vector S is directly connected to the input of every register.

In Figure 4.12 at the first timing point t the leftmost register, register 1, will accept the vector $S(t)$ while the other registers are inhibited. At the next timing point, $t+1$, register 2 will accept the vector $S(t+1)$ while the other registers are inhibited. In this way the next register is always enabled and eventually the whole sequence of vectors $S(t), \ldots, S(t+3)$ is captured and will be available simultaneously. The D flip-flop chain in Figure 4.12 provides the travelling clock pulse for the actual registers, clocking first the first register, then the second, and so on. In this method the first, second, etc., vector will always settle in spatially constant positions. This

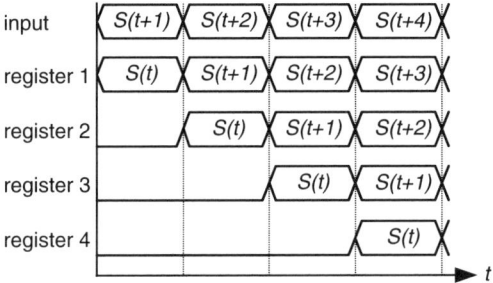

Figure 4.11 The content flow in the shift register chain of Figure 4.10

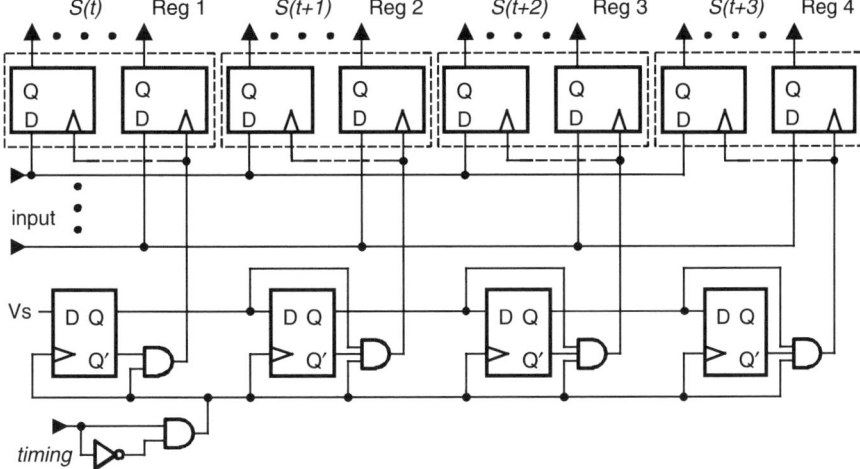

Figure 4.12 Parallel registers for the serial-to-parallel transformation by the parallel method

56 CIRCUIT ASSEMBLIES

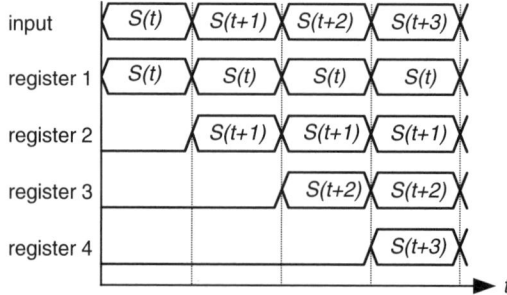

Figure 4.13 The content flow in the registers of Figure 4.12

is presented in Figure 4.13. It can be seen that in this method only a limited length sequence can be accepted and the registers must be reset to the blank state at the beginning of each new input sequence.

It should be obvious that other circuit solutions for the serial-to-parallel transformation may also be devised, as well as some that are more 'neuron-like'.

4.9 PARALLEL-TO-SERIAL TRANSFORMATION

The parallel-to-serial transformation transforms a number of simultaneously available vectors into a temporal sequence:

$$\{S1, S2, \ldots, Sn\} \rightarrow S(t+1) = S1, \ S(t+2) = S2, \ldots, S(t+n) = Sn$$

This transform can be executed by the circuitry of Figure 4.14.

In Figure 4.14 the registers 1, 2, 3 and 4 hold the simultaneously available vectors $S1$, $S2$, $S3$ and $S4$. Each register is connected to the common output bus via the corresponding switch groups $SW1$, $SW2$, $SW3$ and $SW4$. This bus contains separate lines for each signal $s(i)$ of the S vector. At the time point $t+1$ the switch group $SW1$ closes and forwards the vector $S1$ to the output, making the output

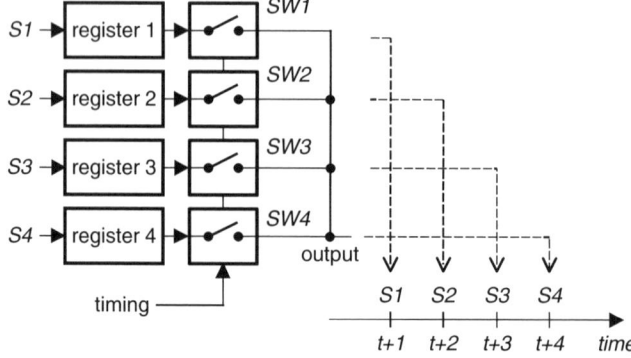

Figure 4.14 The parallel-to-serial transformation of vectors

vector $S(t+1)$ equal to $S1$; at the time point $t+2$ the switch group $SW1$ opens and the switch group $SW2$ closes and forwards the vector $S2$ to the output, making the output vector $S(t+2)$ equal to $S2$; and so on until the last vector $S4$. In this way the vectors $S1$, $S2$, $S3$ and $S4$ will be presented as a temporal sequence.

4.10 ASSOCIATIVE PREDICTORS AND SEQUENCERS

Associative sequences are temporal successions of signal vectors like $S(t)$, $S(t+1)$, $S(t+2)$, ..., where each $S(t+n)$ is associatively connected with one or more previous vectors in the sequence.

An associative predictor circuit predicts the next vector in a sequence when a number of previous vectors are given. An associative sequencer circuit replays an associative sequence in correct order if suitable cues are given. The precondition for prediction and replay is the previous learning of similar sequences. An associative predictor circuit is given in Figure 4.15.

The associative neuron group of Figure 4.15 has the partitioned associative inputs $A1$, $A2$, $A3$ and $A4$. The shift registers 1, 2, 3 and 4 perform the serial-to-parallel transform of the input sequence $S(t)$ so that the four previous S vectors always appear simultaneously at the $A1$, $A2$, $A3$ and $A4$ inputs.

The predictor circuit operates during learning as follows. If a sequence begins at the time point t with $S(t)$ as the input vector, then at the next time point $t+1$ the register 1 has $S(t)$ as its output and as the $A1$ input. At this moment the previous input vector $S(t)$ is associated with the present input vector $S(t+1)$. At the next moment, $t = t+2$, the register 1 has the vector $S(t+1)$ as its output and $A1$ input and the register 2 has the vector $S(t)$ as its output and $A2$ input. At that moment the vectors $S(t+1)$ and $S(t)$ are associated with the present input vector $S(t+2)$. At the next moment the process continues in a similar way (Figure 4.16).

The predictor circuit begins to predict whenever a learned sequence enters it. The first vector $S(t)$ of a learned sequence will evoke the next vector $S(t+1)$, the first and second vectors $S(t)$ and $S(t+1)$ will evoke the next vector $S(t+2)$ and so on. The predictor circuit is able to predict vectors only as long as there are signals at the associative inputs. As the associative input signals are delayed versions of the main input signal S, the prediction output will go to zero at the latest after the total delay of the shift register chain.

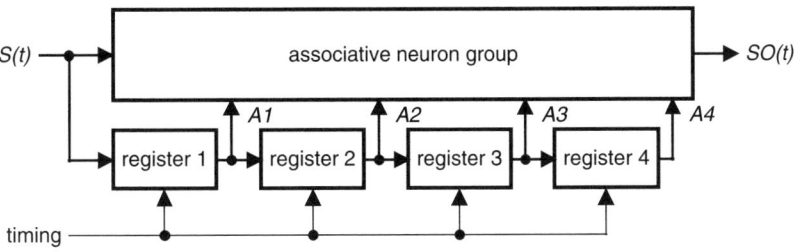

Figure 4.15 An associative predictor circuit

58 CIRCUIT ASSEMBLIES

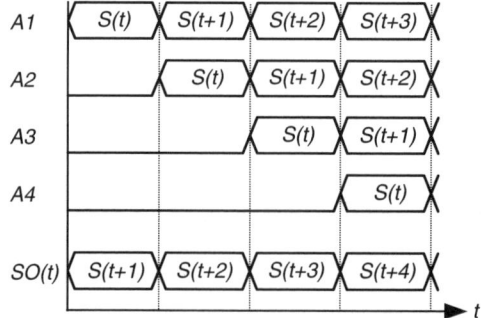

Figure 4.16 The register contents and output in the predictor circuit

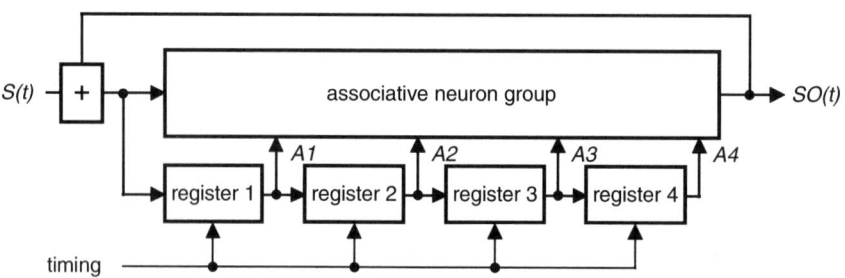

Figure 4.17 An associative predictor/sequencer circuit

Predictor/sequencer circuits are able to produce sequences of indefinite length. An associative predictor circuit may be transformed into an associative predictor/sequencer circuit by looping the output from the predictor neuron group back to the input (Figure 4.17).

A learned sequence may be evoked by inputting the first few $S(t)$ vectors of the sequence. The evocation will continue until the end of the sequence as the evoked continuation is relayed via the register network to the associative inputs of the associative neuron group.

These predictor and predictor/sequencer circuits are autoassociative; they use parts of their learned sequences as the evocation cues. They also suffer from an initiation branching problem as initially there is only one cue vector, the first vector $S(t)$ of the sequence that is used to evoke the continuation. This may lead to ambiguity as there may be several different sequences that begin with the same vector $S(t)$, especially if the vector $S(t)$ is short. Consequently, the circuit has no way of knowing which sequence is intended and a prediction error may occur.

The command-to-sequence circuit remedies the branching problem by using additional information. The command-to-sequence circuit of Figure 4.18 uses a command input vector C as the cue and as the additional information.

When the command-to-sequence circuit of Figure 4.18 begins to learn, the first S input vector $S(t)$ in a sequence will be associated with the command vector only ($A3 = C$). The second S input vector $S(t+1)$ will be associated with the previous S input vector $S(t)$ ($A1 = S(t)$) and with the command vector as the $A4$ input

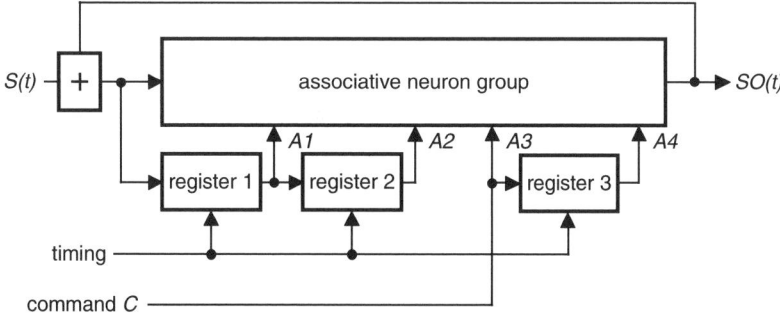

Figure 4.18 The command-to-sequence circuit

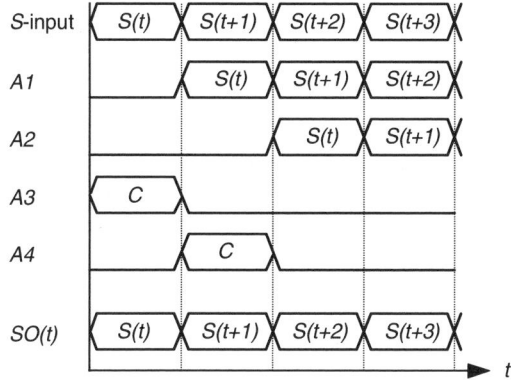

Figure 4.19 The register contents and association in the command-to-sequence circuit

($A4 = C$). The next S input vector $S(t+2)$ is associated with the vectors $S(t+1)$ and $S(t)$. The register content flow of the command-to-sequence circuit is shown in Figure 4.19.

The replay of a sequence begins with the introduction of the command vector C. This evokes the first vector of the sequence, $SO(t) = S(t)$, which is looped back to the S input. At the next moment this S input vector $S(t)$ is shifted through the register 1 and then becomes the $A1$ input. The C vector is also shifted through the register 3 and becomes the $A4$ input. These will now evoke the next vector in the sequence, $SO(t+1) = S(t+1)$. The continuation of the operation should be seen from the diagram of Figure 4.19.

The label-sequence circuit associates a sequence with and evokes it by only one vector, 'a label'. In Figure 4.20 the static 'label' vector A is to be associated with a sequence $S(t)$. Here synaptic partitioning is used with timed switching. At the time point $t = 1$ the vector A is associated with the first vector $S(1)$ of the sequence, at the time point $t = 2$ the vector A is associated with the second vector $S(2)$ of the sequence and so on, each time utilizing different synaptic segments.

During replay the switch $SW1$ is closed first while the other switches remain open and the output $SO(1)$ is evoked. Then the switch $SW1$ is opened and the switch $SW2$ is closed, causing the evocation of $SO(2)$ and so on until the end of the

60 CIRCUIT ASSEMBLIES

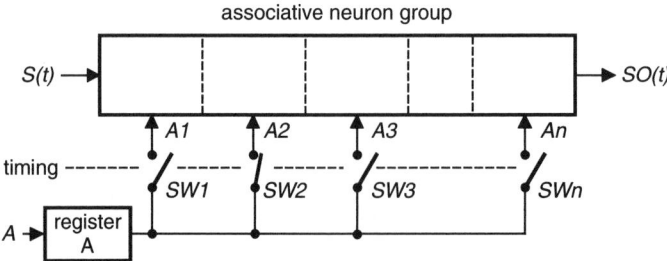

Figure 4.20 The label-sequence circuit; the evocation of a sequence by a 'label' vector A

sequence. This circuit may be used to evoke a sequence of phonemes or letters, a word or name, as a response to a label vector A. The operation of the label-sequence circuit is described in Figure 4.21.

The sequence-label circuit executes the inverse of the label-sequence operation. Here a sequence must be associated with a static vector, a 'label', for the sequence so that the label can be evoked by the exact sequence or a sequence that is similar enough. Let the sequence be $A(t)$ and the label vector S. For the association and evocation operations the sequence $A(t)$ must be transformed into a temporally parallel form. Thereafter the association and evocation can be executed by the associative neuron group with the circuit of Figure 4.22.

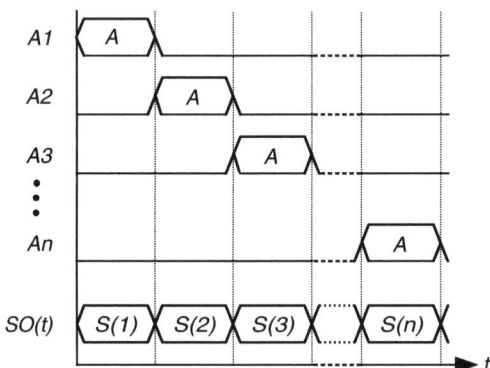

Figure 4.21 The operation of the label-sequence circuit

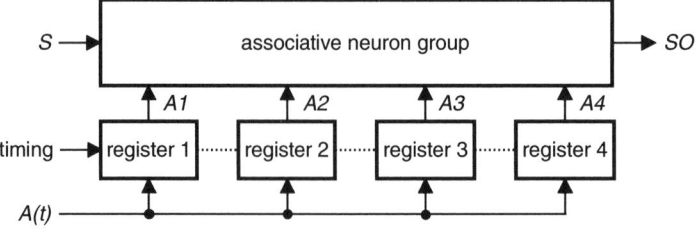

Figure 4.22 The sequence-label circuit

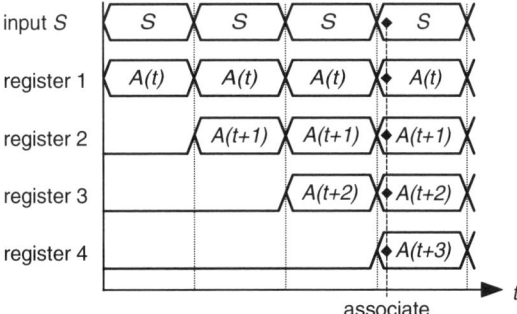

Figure 4.23 The operation of the sequence-label circuit

In Figure 4.22 the registers 1, 2, 3 and 4 operate as a serial-to-parallel transformer in the style of Figures 4.12 and 4.13, and capture the instantaneous vectors from the sequence $A(t)$. The signal flow of the sequence-label circuit is depicted in Figure 4.23.

In Figure 4.23 all the vectors of the sequence are available simultaneously as soon as the last register has captured its vector. Consequently, the association and evocation operations can be executed at that point. In practice there might be a much larger number of these registers, but the operational principle would be the same. The sequence-label circuit can be used to detect and recognize given temporal sequences.

For proper timing these sequence circuits require external timing sources. The principles of sequence timing are discussed in the following.

4.11 TIMING CIRCUITS

Timing circuits are required for the estimation, memorization and reproduction of the duration of the intervals in sequences (for example the rhythm of a melody or the timing of word pronunciation). This should be done in a way that also allows the recognition of rhythms. The possibility of modifying the overall tempo while conserving the proportionality of individual intervals during reproduction would be useful.

In a sequence of vectors each successive vector has its own temporal duration. Thus the reproduction of a sequence calls for the evocation of each vector and its temporal duration in the correct order. Here the temporal presence of a vector in a sequence is called an event. Events are characterized by their corresponding vectors and temporal durations.

In Figure 4.24 the event $e1$ corresponds to the temporal presence of the vector $S(e1) = \{1, 0, \ldots, 0\}$. The event $e2$ corresponds to the vector $S(e2) = \{0, 0, \ldots, 0\}$, which is a zero vector corresponding to a pause. The event $e3$ corresponds to the vector $S(e3) = \{0, 0, \ldots, 1\}$, etc.

The duration of the event ei is defined here as the temporal duration of the corresponding signal vector $S(ei) = \{s(0), s(1), \ldots, s(n)\}$. In order to determine the

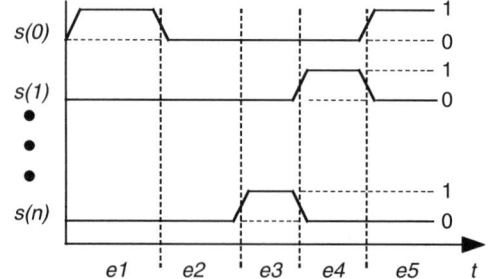

Figure 4.24 Events of the sequence of the vector $S(e)$

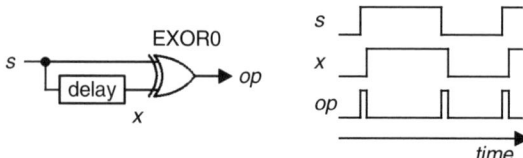

Figure 4.25 A transition detector

duration of events the beginning and end time points of the events must be detected. The end of one event and the beginning of another is marked by a transition from zero to one or from one to zero in any of the individual signals $s(0)$, $s(1)$, ..., $s(n)$ of the signal vector $S(ei)$.

The transitions from zero to one and from one to zero in a signal s may be detected by the circuit of Figure 4.25. The transition detector circuit of Figure 4.25 executes the logical function $op = s$ EXOR x, where x is the delayed version of the signal s. Due to the delay the s and x signals are unequal for the period of delay after each transition from zero to one and from one to zero. The EXOR circuit gives the logical one output whenever its inputs are unequal and therefore the output op will indicate the moments of transition.

A signal vector S consists of a number of individual signals $s(0)$, $s(1)$, ..., $s(n)$. The transitions that mark the beginnings and ends of events may occur in any of these signals. Therefore each individual signal must have its own transition detector and the outputs of these detectors must be combined into one output signal, which will be called here the 'reset/start pulse'. This can be done using the circuit of Figure 4.26. The outputs of the individual transition detector circuits are combined by a logical OR circuit. Thus a change in any of the input signals will cause a final output pulse, the reset/start pulse.

The reproduction of timed sequences call for the measurement or estimation of the duration of each event in the sequence. Associative neuron groups operate with signal vectors; therefore it would be useful if the event durations could also be represented by signal vectors. This would allow the representation of an event by two cross-associated signal vectors, namely the actual vector S ('what happens') and the corresponding event duration vector D ('how long'). In the following one such method is presented.

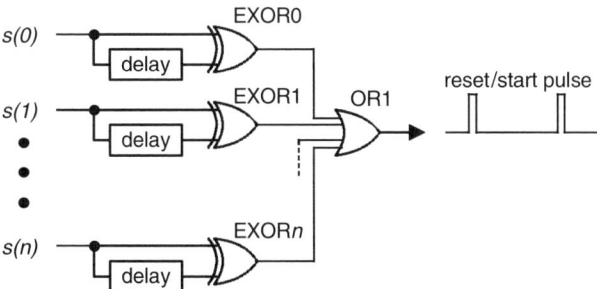

Figure 4.26 A change detector; the extraction of the event begin/end time points from a sequence of vectors

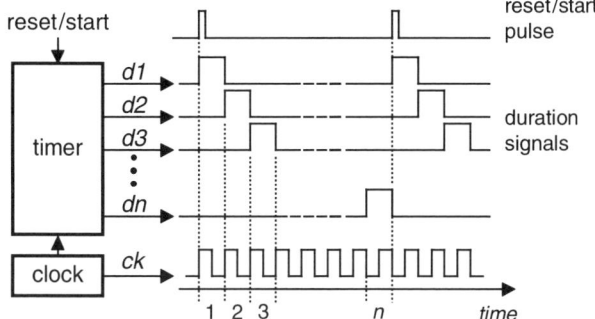

Figure 4.27 The timing of an event with duration signals

The event duration is the time interval between successive transitions in the S vector, which can be indicated by the reset/start pulse from the change detector output of Figure 4.26. Figure 4.27 depicts a circuit that can determine the event duration using a timer device that is controlled by the reset/start pulses.

In Figure 4.27 a timer circuit generates a time-varying duration signal vector $D(t) = \{d0, d1, d2, \ldots, dn\}$ that indicates the length of the elapsed time from the timer start moment. Only one of these signals $\{d0, d1, d2, \ldots, dn\}$ is active at a time and has a fixed short duration Δt. This duration determines the timing resolution of the event. The duration of an event, that is the time between successive reset/start pulses, is $n^*\Delta t$, where n is the number of the last duration signal.

4.12 TIMED SEQUENCE CIRCUITS

In order to illustrate the requirements of sequence timing a circuit that learns and reproduces timed sequences is presented in Figure 4.28. This circuit is based on the combination of the predictor/sequencer circuit of Figure 4.17 and the previously discussed timing circuits.

In Figure 4.28 the associative neuron group 1 and the registers 1, 2 and 3 operate as the predictor/sequencer circuit that learns and predicts S vector input

64 CIRCUIT ASSEMBLIES

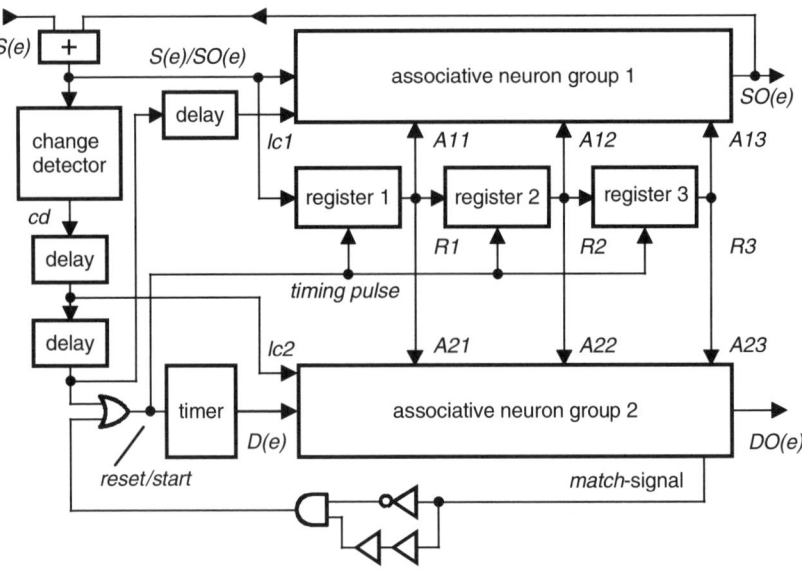

Figure 4.28 A circuit assembly that learns and reproduces timed sequences

sequences. The change detector detects change points in the input sequence and produces corresponding change detection pulses cd. The timer produces a running $D(e)$ timing vector which is reset by delayed cd pulses. The associative neuron group 2 associates the last $D(e)$ vector of each episode with the three previous $S(e)$ vectors that are the outputs $R1$, $R2$ and $R3$ of the registers 1, 2 and 3 respectively. Thus the neuron group 2 will learn the sequences of the interval durations of the S vector sequences. The timing diagram for the operation during learning is given in Figure 4.29.

In Figure 4.28 the timer produces the single signal vector $D(e) = \{d0, d1, d2, \ldots, dn\}$ where the last $di = 1$ indicates the duration of the interval. This di should now be associated with the simultaneous $A21$, $A22$ and $A23$ vectors, which are the three previous $S(e)$ vectors. However, the fact that di is the last duration signal is known only after the end of the event. At that point the signal vectors $A21$, $A22$ and $A23$ and the corresponding $D(e)$ vector (di signal) would no longer be available for the desired association with each other if the shift registers 1, 2 and 3 were timed directly by the change detection cd signal. Therefore the timing of the *set/reset* signal must be delayed so that the association can be made and the learning command signals $lc1$ and $lc2$ must be generated in suitable time points. This introduces a small delay to the operation, but this is insignificant in practice as typically the event durations are of the order of milliseconds or more and the required delays are of the order of microseconds or less. The moment of association and the associated signals are depicted by the symbol ♦ in Figure 4.29.

The replay begins with the introduction of the first few $S(e)$ vectors of the sequence to be replayed. When these enter the shift register chain they will evoke the vector $SO(e)$ at the neuron group 1 output, which will be the same as the

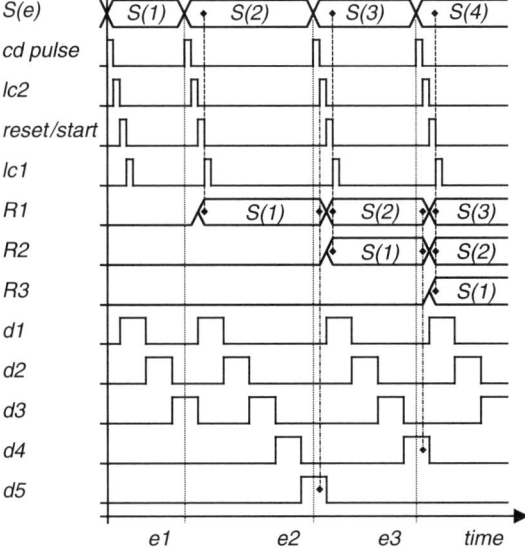

Figure 4.29 The timing during learning

instantaneous input vector $S(e)$. At the neuron group 2 output the corresponding duration vector $DO(e)$ is evoked. The timer will run from zero to the point where the $D(e)$ vector matches this $DO(e)$ vector. This elapsed time corresponds to the original duration of the $S(e)$ vector (within the timer's resolution). The neural *match* signal is generated during the matching *di* signal. However, the episode is supposed to end only at the end of the *di* signal and consequently at the end of the *match* signal. Therefore additional circuitry is required that generates a short pulse at the trailing edge of the *match* signal. This signal can be used as the reset/start pulse for the timer and also as the timing pulse for the shift registers 1, 2 and 3. As the output vector $SO(e)$ is looped back to the input, the initiated sequence will continue even if the actual input is removed. The timing during replay is presented in Figure 4.30.

This method has the following benefits. The temporal event durations are represented by signal vectors, which can be associatively handled. The replay speed can be changed while conserving the proportionality of the individual event durations by changing the timer clock speed.

It is obvious that there are also other ways to realize the functionality of the timed sequence circuit. Therefore a general sequence neuron assembly is defined here which is able to learn and replay timed input sequences and, in addition, has an associative command input that can evoke learned and named sequences (see Figure 4.18). The symbol for the sequence neuron assembly is presented in Figure 4.31.

In Figure 4.31 $S(e)$ is the sequence input, $SO(e)$ is the evoked sequence output and *command* C is the associative command input. The timing is assumed to take place inside the neuron assembly. A learned sequence may be initiated by the $S(e)$

66 CIRCUIT ASSEMBLIES

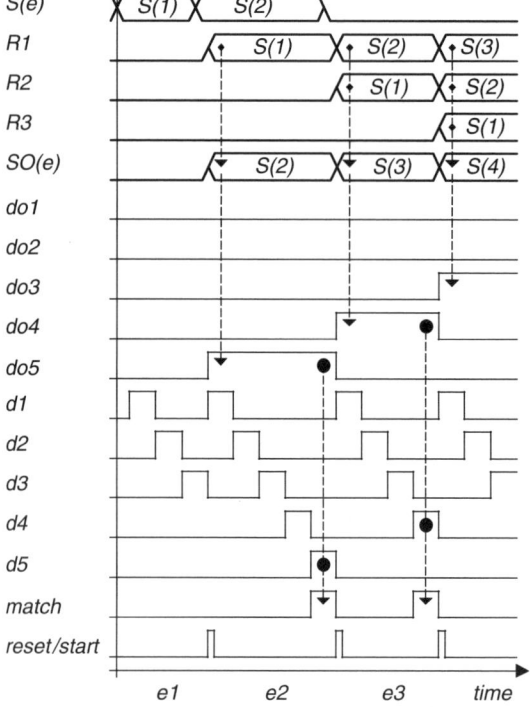

Figure 4.30 The timing during replay

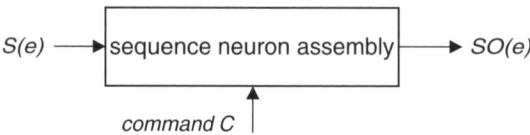

Figure 4.31 The sequence neuron assembly symbol

vectors or by the command C vector. The sequence neuron assembly can be realized as a short-term memory or as a long-term memory.

4.13 CHANGE DIRECTION DETECTION

In single signal representation vectors like $\{p(0), p(1), p(2), p(3), p(4)\}$ only one $p(i)$ may be nonzero at any given time. If this vector represents, for instance, a position then any nonzero $p(i)$ represents the instantaneous position of the related object. Accordingly the motion of the object is seen as a sequential change of value of the $p(i)$ signals, as shown in Figure 4.32.

The direction of the motion can be deduced from a pair of signals $p(i)$, $p(i+1)$ at the moment of transition, when one of these signals goes from *one* to *zero* and the other from *zero* to *one*. Change direction can be detected by the circuit of Figure 4.33.

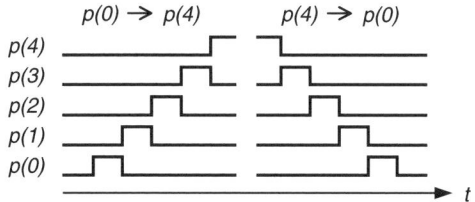

Figure 4.32 Motion in a position vector

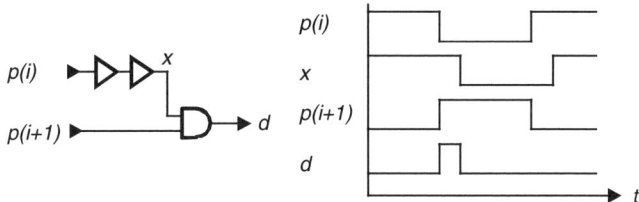

Figure 4.33 A change direction detector for a signal pair

The circuit of Figure 4.33 detects the situation when the signal $p(i)$ goes to *zero* and the signal $p(i+1)$ goes to *one*. For this purpose a delayed version x of the signal $p(i)$ is produced and the logical function $d = x$ AND $p(i+1)$ is formed. It can be seen that $d = 1$ immediately after the transition $p(i) \rightarrow 0$, $p(i+1) \rightarrow 1$. On the other hand, the transition $p(i) \rightarrow 1$, $p(i+1) \rightarrow 0$ does not have any effect on d. Thus this circuit is able to detect the case when the motion direction is from position $p(i)$ to position $p(i+1)$.

The circuit of Figure 4.33 can be applied to a larger circuit that is able to detect motion in both directions over the whole vector $\{p(0), p(1), p(2), \ldots, p(n)\}$ (Figure 4.34). In Figure 4.34 the upper part of the circuit detects the motion direction from the position $p(0)$ towards the position $p(n)$. The detection is performed for each signal pair separately and the final detection result is achieved by the logical

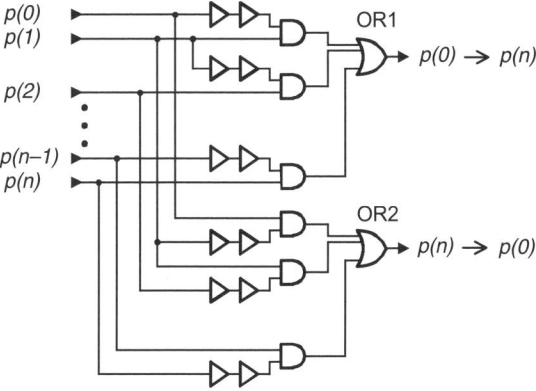

Figure 4.34 A bidirectional motion detector

OR operation (OR1). Thus the output will be *one* whenever the corresponding change direction is detected by any of the signal pair detectors. The lower part of the circuit detects the motion direction from the position $p(n)$ towards the position $p(0)$ in a similar way.

These circuits work well for ideal signals. In practice the effects of noise and imperfect signal forms must be considered.

5
Machine perception

5.1 GENERAL PRINCIPLES

Machine perception is defined here as the process that allows the machine to access and interpret sensory information and introspect its own mental content. Sensory perception is seen to involve more than the simple reception of sensory data. Instead, perception is considered as an active and explorative process that combines information from several sensory and motor modalities and also from memories, models, in order to make sense and remove ambiguity. These processes would later on enable the imagination of the explorative actions and the information that might be revealed if the actions were actually executed. There seems to be experimental proof that also in human perception explorative actions are used in the interpretation of percepts (Taylor, 1999; O'Regan and Noë, 2001; Gregory, 2004, pp. 212–218).

Thus, the interpretation of sensory information is not a simple act of recognition and labelling; instead it is a wider process involving exploration, expectation and prediction, context and the instantaneous state of the system. The perception process does not produce representations of strictly categorized objects; instead it produces representations that the cognitive process may associate with various possibilities for action afforded by the environment. This view is somewhat similar to that of Gibson (1966). However, perception is not only about discrete entities; it also allows the creation of mental scenes and maps of surroundings – what is where and what would it take to reach it? Thus, perception, recognition and cognition are intertwined and this classification should be seen only as a helping device in this book.

In the context of conscious machines the perception process has the additional requisite of transparency. Humans and obviously also animals perceive the world lucidly, apparently without any perception of the underlying material processes. (It should go without saying that the world in reality is not necessarily in the way that our senses present it to us.) Thoughts and feelings would appear to be immaterial and this observation leads to the philosophical mind–body problem: how a material brain can cause an immaterial mind and how an immaterial mind can control the material brain and body. The proposed solution is that the mind is not immaterial at all; the apparent effect of immateriality arises from the transparency of the carrying material processes (Haikonen, 2003a). The biological neural system stays transparent and only the actual information matters. Transparent systems are known in technology, for example the modulation on a radio carrier wave; it is the program that is heard, not the carrier wave. This as well as the transistors and other circuitry of the radio set remain transparent (see also Section 11.2, 'Machine perception and qualia', in Chapter 11).

Traditional computers and robots utilize digital signal processing, where sensory information is digitized and represented by binary numbers. These numeric values are then processed by signal processing algorithms. It could be strongly suspected that this kind of process does not provide the desired transparent path to the system and consequently does not lead to lucid perception of the world. Here another approach is outlined, one that aims at transparent perception by using distributed signal representations and associative neuron groups. The basic realization of the various sensory modalities is discussed with reference to human perception processes where relevant.

5.2 PERCEPTION AND RECOGNITION

Traditional signal processing does not usually make any difference between perception and recognition. What is recognized is also perceived, the latter word being redundant and perhaps nontechnical. Therefore the problem of perception is seen as the problem of classification and recognition. Traditionally there have been three basic approaches to pattern recognition, namely the template matching methods, the feature detection methods and the neural network methods.

In the template matching method the sensory signal pattern is first normalized and then matched against all templates in the system's memory. The best-matching template is taken as the recognized entity. In vision the normalization operation could consist, for example, of rescaling and rotation so that the visual pattern would correspond to a standard outline size and orientation. Also the luminance and contrast values of the visual pattern could be normalized. Thereafter the visual pattern could be matched pixel by pixel against the templates in the memory. In the auditory domain the intensity and tempo of the sound patterns could be normalized before the template matching operation. Template matching methods are practical when the number of patterns to be recognized is small and these patterns are well defined.

The feature detection method is based on structuralism, the idea that a number of detected features (sensations) add up to the creation of a percept of an object. (This idea was originally proposed by Wilhelm Wundt in about 1879.) The Pandemonium model of Oliver Selfridge (1959) takes this idea further. The Pandemonium model consists of hierarchical groups of detectors, 'demons'. In the first group each demon (a feature demon) detects its own feature and 'shouts' if this feature is present. In the second group each demon (a cognitive demon) detects its own pattern of the shouting feature demons of the first group. Finally, a 'decision demon' detects the cognitive demon that is shouting the loudest; the pattern that is represented by this cognitive demon is then deemed to be the correct one. Thus the pattern will be recognized via the combination of the detected features when the constituting features are detected imperfectly.

Feature detection methods can be applied to different sensory modalities, such as sound and vision. Examples of visual object recognition methods that involve elementary feature detection and combination are David Marr's (1982) computational approach, Irving Biederman's (1987) recognition by components (RBC) and Anne Treisman's (1998) feature integration theory (FIT).

There are a number of different neural network based classifiers and recognizers. Typically these are statistically trained using a large number of examples in the hope of getting a suitable array of synaptic weights, which would eventually enable the recognition of the training examples plus similar new ones. The idea of structuralism may be hidden in many of these realizations.

Unfortunately, none of these methods guarantee flawless performance for all cases. This is due to fundamental reasons and not to any imperfections of the methods.

The first fundamental reason is that in many cases exactly the same sensory patterns may depict completely different things. Thus the recognition by the properties of the stimuli will fail and the true meaning may only be inferred from the context. Traditional signal processing has recognized this and there are a number of methods, like the 'hidden Markow processes', that try to remedy the situation by the introduction of a statistical context. However, this is not how humans do it. Instead of some statistical context humans utilize 'meaning context' and inner models. This context may partly reside in the environment of the stimuli and partly in the 'mind' of the perceiver. The existence of inner models manifests itself, for instance, in the effect of illusory contours.

The second fundamental reason is that there are cases when no true features are perceived at all, yet the depicted entity must be recognized. In these cases obviously no template matching or feature detecting recognizers can succeed. An example of this kind of situation is the pantomime artist who creates illusions of objects that are not there. Humans can cope with this, and true cognitive machines should also do so.

Thus recognition is not the same as perception. Instead, it is an associative and cognitive interpretation of percepts calling for feedback from the system. It could even be said that humans do not actually recognize anything; sensory percepts are only a reminder of something. The context-based recognition is known in cognitive psychology as 'conceptually driven recognition' or 'top–down processing'. This approach is pursued here with feature detection, inner models and associative processing and is outlined in the following chapters.

5.3 SENSORS AND PREPROCESSES

A cognitive system may utilize sensors such as microphones, image sensors, touch sensors, etc., to acquire information about the world and the system itself. Usually the sensor output signal is not suitable for associative processing, instead it may be

72 MACHINE PERCEPTION

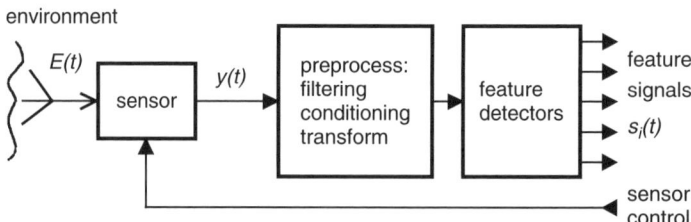

Figure 5.1 A sensor with preprocess and feature detection

in the form of raw information containing the sum and combination of a number of stimuli and noise. Therefore, in order to facilitate the separation of the stimuli some initial preprocessing is needed. This preprocess may contain signal conditioning, filtering, noise reduction, signal transformation and other processing. The output of the preprocess should be further processed by an array of feature detectors. Each feature detector detects the presence of its specific feature and outputs a signal if that feature is present. The output of the feature detector array is a distributed signal vector where each individual signal carries one specific fraction of the sensed information. Preferably these signals are orthogonal; a change in the fraction of information carried by one signal should not affect the other signals.

Figure 5.1 depicts the basic steps of sensory information acquisition. $E(t)$ is the sum and combination of environmental stimuli that reach the sensor. The sensor transforms these stimuli into an electric output signal $y(t)$. This signal is subjected to preprocesses that depend on the specific information that is to be processed. Preprocesses are different for visual, auditory, haptic, etc., sensors. Specific features are detected from the preprocessed sensory signal.

In this context there are two tasks for the preprocess:

1. The preprocess should generate a number of output signals $s_i(t)$ that would allow the detection and representation of the entities that are represented by the sensed stimuli.

2. The output signals should be generated in the form of feature signals having the values of a positive value and zero, where a positive value indicates the presence of the designated feature and zero indicates that the designated feature is not present. Significant information may be modulated on the feature signal intensities.

5.4 PERCEPTION CIRCUITS; THE PERCEPTION/RESPONSE FEEDBACK LOOP

5.4.1 The perception of a single feature

The perception of a single feature is the simplest possible perceptual task. Assume that the presence or absence of a feature is detected by a sensory feature detector and the result of this detection is represented by one on/off (binary) signal. Perception

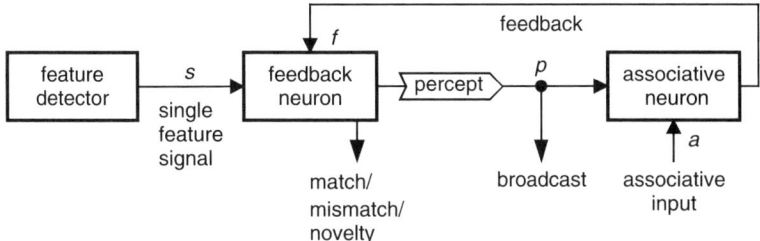

Figure 5.2 The single signal perception/response feedback loop

is not the simple reception of a signal; instead the sensed signal or its absence should be assessed against the internal state of the cognitive system. A simple circuit module that can do this is presented in Figure 5.2. This module is called the single signal perception/response feedback loop.

In Figure 5.2 a feature detector outputs one binary signal s that indicates the detected presence or absence of the corresponding feature; the intrinsic meaning of this signal is the detected feature. The signal s is forwarded to the main signal input of the so-called feedback neuron. The associative input to the feedback neuron is the signal f, which is the feedback from the system. The intrinsic meaning of the feedback signal f is the same as that of the feature signal s. The feedback neuron combines the effects of the feature signal s and the feedback signal f and also detects the match, mismatch and novelty conditions between these. The percept arrow box is just a label that depicts the point where the 'official' percept signal p is available. The percept signal p is forwarded to the associative neuron and is also broadcast to the rest of the system.

The intensity of the percept signal is a function of the detected feature signal and the feedback signal and is determined as follows:

$$p = k_1^* s + k_2^* f \qquad (5.1)$$

where

$p =$ percept signal intensity
$s =$ detected feature signal intensity, binary
$f =$ feedback signal intensity, binary
$k_1 =$ coefficient
$k_2 =$ coefficient

Note that this is a simplified introductory case. In practical applications the s and f signals may have continuous values.

The match, mismatch and novelty indicating signal intensities are determined as follows:

$$m = s^* f \qquad \text{(match condition)} \qquad (5.2)$$

$$mm = f^* (1-s) \qquad \text{(mismatch condition)} \qquad (5.3)$$

74   MACHINE PERCEPTION

$$n = s*(1-f) \qquad \text{(novelty condition)} \qquad (5.4)$$

where

m = match signal intensity, binary
mm = mismatch signal intensity, binary
n = novelty signal intensity, binary

The associative neuron associates the percept p with an associative input signal a. Therefore, after this association the signal a may evoke the feedback signal f, which, due to the association, has the same meaning as the percept signal p. The feedback signal f may signify an expected or predicted occurrence of the feature signal s or it may just reflect the inner states of the system.

Four different cases can be identified:

$$s=0, f=0 \Rightarrow p=0, m=0, mm=0, n=0$$

Nothing perceived, nothing expected, the system rests.

$$s=1, f=0 \Rightarrow p=k_1, m=0, mm=0, n=1$$

The feature s is perceived, but not expected. This is a novelty condition.

$$s=0, f=1 \Rightarrow p=k_2, m=0, mm=1, n=0$$

The feature s is predicted or expected, but not present. The system may be searching for the entity with the feature s and the search at that moment is unsuccessful. This is a mismatch condition.

$$s=1, f=1 \Rightarrow p=k_1+k_2, m=1, mm=0, n=0$$

The feature s is expected and present. This is a match condition. If the system has been searching for an entity with the feature s then the search has been successful and the match signal indicates that the searched entity has been found.

The feedback signal f can also be understood as a priming factor that helps to perceive the expected features. According to Equation (5.1) the intensity of the percept signal p is higher when both $s=1$ and $f=1$. This higher value may pass possible threshold circuits more easily and would thus be more easily accepted by the other circuits of the system.

5.4.2 The dynamic behaviour of the perception/response feedback loop

Next the dynamic behaviour of the perception/response feedback loop is considered with a simplified zero delay perception/response feedback loop model (Figure 5.3).

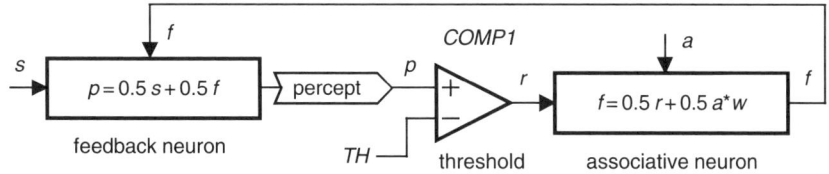

Figure 5.3 A single signal perception/response zero delay feedback loop model

It is assumed that the feedback loop operates without delay and the associative neuron produces its output immediately. However, in practical circuits some delay is desirable.

In Figure 5.3 the associative neuron has an input comparator COMP1. This comparator realizes a limiting threshold function with the threshold value TH. The markings in Figure 5.3 are:

s = input signal intensity
p = percept signal intensity
r = threshold comparator output signal intensity, 1 or 0
f = feedback signal intensity
a = associative input signal intensity, 1 or 0
TH = threshold value
w = synaptic weight value; $w = 1$, when r and a are associated with each other

Note that in certain applications the signals s and a may have continuous values, in this example only the values 0 and 1 are considered.

The input threshold for the associative neuron is defined as follows:

$$\text{IF } p > TH \text{ THEN } r = 1 \text{ ELSE } r = 0 \tag{5.5}$$

According to Equation (5.3) the percept signal intensity will be

$$p = 0.5*s + 0.5*f = 0.5*s + 0.25*r + 0.25*a \tag{5.6}$$

In Figure 5.4 the percept signal intensity p is depicted for the combinations of the s and a intensities and the threshold values $TH = 0.2$ and $TH = 0.8$. It can be seen that normally the percept signal level is zero. An active s signal with the value 1 will generate a percept signal p with the value of 0.75. If the s signal is removed, the percept signal p will not go to zero; instead it will remain at the lower level of 0.25. This is due to the feedback and the nonlinear amplification that is provided by the threshold comparator; the signal reverberates in the feedback loop. Similar reverberation takes place also for the associative input signal a. In this way the perception/response feedback loop can operate as a short-term memory. The reverberation time can be limited by, for instance, AC coupling in the feedback line.

When the threshold TH is raised to the value of 0.8 the comparator output value goes to zero, $r = 0$, and the feedback loop opens and terminates any ongoing reverberation. Consequently the high threshold value will lower the percept intensities

76 MACHINE PERCEPTION

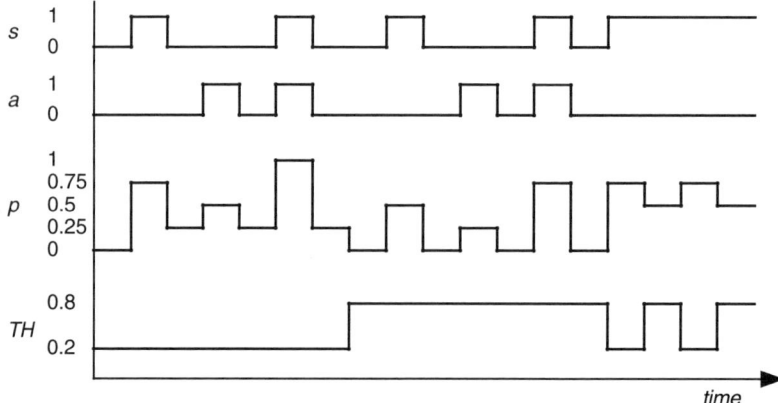

Figure 5.4 The percept signal p intensity in the perception/response loop

Table 5.1 The behaviour of the perception/response loop

TH	s	a	p	Comments
0.2	0	0	0	
	0	0	0.25	Reverberation, provided that s or a has been 1
	0	1	0.5	Introspective perception
	1	0	0.75	Sensory perception without priming
	1	1	1	Sensory perception with priming
0.8	0	0	0	No reverberation
	0	1	0.25	Introspective perception
	1	0	0.5	Sensory perception without priming
	1	1	0.75	Sensory perception with priming

in each case. In this way the threshold value can be used to modulate the percept intensity. The behaviour of the perception/response loop is summarized in Table 5.1.

5.4.3 Selection of signals

In a typical application a number of perception/response feedback loops broadcast their percept signals to the associative inputs of an auxiliary neuron or neurons. In this application the system should be able to select the specific perception/response loop whose percept signal would be accepted by the receiving neuron. This selection can be realized by the associative input signal with the circuit of Figure 5.5.

In Figure 5.5 the auxiliary neuron has three associative inputs with input threshold comparators. The associative input signals to the auxiliary neuron are the percept signals $p1$, $p2$ and $p3$ from three corresponding perception/response loops. If the threshold value for these comparators is set to be 0.8 then only percept signals with intensities that exceed this will be accepted by the comparators. Previously it was shown that the percept signal p will have the value 1 if the signals s and

PERCEPTION CIRCUITS   77

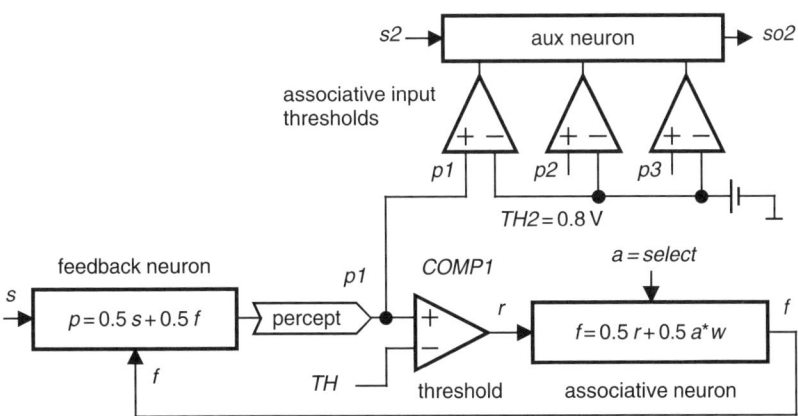

Figure 5.5 The associative input signal as the selector for the perception/response loop

a are 1 simultaneously, otherwise the percept signal will have the value of 0.75 or less. Thus the signal *a* can be used as a selector signal that selects the desired perception/response loop. In practical applications, instead of one *a* signal there is a signal vector *A* and in that case the selection takes place if the signal vector *A* has been associated with the *s* signal of the specific perception/response loop.

5.4.4 Perception/response feedback loops for vectors

Real-world entities have numerous features to be detected and consequently the perception process would consist of a large number of parallel single signal perception/response loops. Therefore the single signal feedback neuron and the associative neuron should be replaced by neuron groups (Figure 5.6).

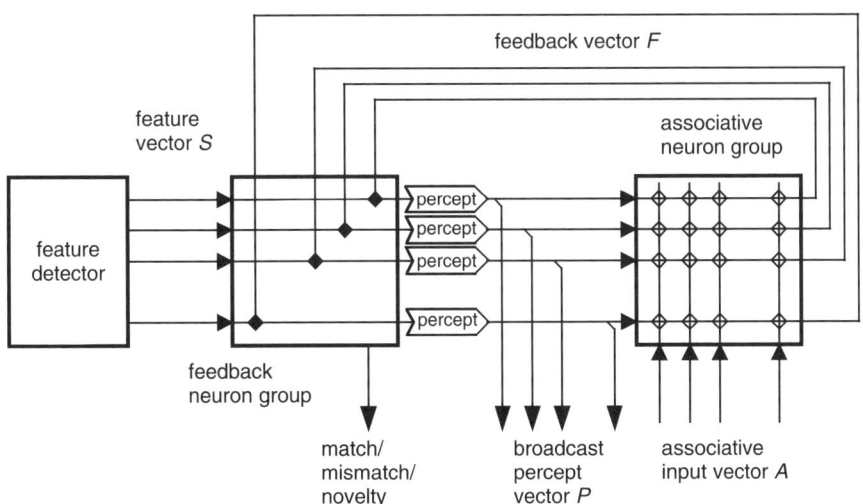

Figure 5.6 The perception/response loop for signal vectors

In Figure 5.6 the feedback neuron group forms a synaptic matrix of the size $k \times k$. The feedback neuron group is a degenerated form of the general associative neuron group with fixed connections and simplified match/mismatch/novelty detection. The synaptic weight values are

$$w(i, j) = 1 \text{ IF } i = j, \text{ ELSE } w(i, j) = 0 \tag{5.7}$$

Thus in the absence of the S vector the evoked vector P will be the same as the evoking feedback vector F.

The vector match M, vector mismatch MM and vector novelty N conditions between the S vector and the F vector are determined at the feedback neuron group. These values are deduced from the individual signal match m, mismatch mm and novelty n values. These are derived as described before. The Hamming distance between the S vector and the F vector can be computed for the feedback neuron group as follows:

$$mm(i) + n(i) = s(i) \text{ EXOR } f(i) \tag{5.8}$$
$$Hd = \Sigma(mm(i) + n(i))$$

where

Hd = Hamming distance between the S vector and the F vector

Thus the vector match M between the S vector and the F vector can be defined as follows:

$$\text{IF } \Sigma(mm(i) + n(i)) < threshold \text{ THEN } M = 1 \text{ ELSE } M = 0 \tag{5.9}$$

where *threshold* determines the maximum number of allowable differences. The vector mismatch MM may be determined as follows:

$$\text{IF } M = 0 \text{ AND } \Sigma mm(i) \geq \Sigma n(i) \text{ THEN } MM = 1 \text{ ELSE } MM = 0 \tag{5.10}$$

The vector novelty N may be determined as follows:

$$\text{IF } M = 0 \text{ AND } \Sigma mm(i) < \Sigma n(i) \text{ THEN } N = 1 \text{ ELSE } N = 0 \tag{5.11}$$

In Figure 5.5 the associative input vector A originates from the rest of the system and may represent completely different entities. The associative neuron group associates the vector A with the corresponding feedback vector F so that later on the vector A will evoke the vector F. The vector F is fed back to the feedback neuron group where it evokes a signal-wise similar percept vector P. If no sensory vector S is present, then the resulting percept vector P equals the feedback vector F. On the other hand, every signal in each percept vector has an intrinsic meaning that is grounded to the point-of-origin feature detector and, accordingly, each percept vector therefore represents a combination of these features. The feedback vector evokes a percept of an entity of the specific sensory modality; in the visual modality the percept is that of a visual entity, in the auditory modality the percept is that of a sound, etc. Thus the perception/response loop transforms the A vector into the equivalent of a sensory percept. In this way the inner content of the system is made available as a percept; the system is able to introspect. This kind of introspection in the visual modality would correspond to visual imagination and in the auditory modality the introspection would appear as sounds and especially as inner speech. This inner speech would appear as 'heard speech' in the same way as humans perceive their own inner speech. However, introspective percepts do not necessarily have all the attributes and features that a real sensory percept would have; introspective imagery may not be as 'vivid'.

It is desired that during imagination the internally evoked percepts would win the externally evoked percepts. This can be achieved by the attenuation of the S vector signals. The S vector signals should be linearly attenuated instead of being completely cut off, so that the S-related attenuated percepts could still reach some of the other modalities of the system (such as the emotional evaluation, see Chapter 8).

5.4.5 The perception/response feedback loop as predictor

The perception/response loop has another useful property, which is described here in terms of visual perception. If a percept signal vector is looped through the associative neuron group and back to the feedback neurons, a self-sustaining closed loop will occur and the percept will sustain itself for a while even when the sensory input changes or is removed; a short-term memory function results. Thus it can be seen that the feedback supplies the feedback neurons with the previous value of the sensory input. This feedback can be considered as the *first-order prediction* for the sensory input as the previous sensory input is usually a good prediction for the next input. If the original sensory input is still present then the match condition will occur; if the sensory input has changed then the mismatch condition will occur.

Consider two crossing bars (Figure 5.7). Assume that the gaze scans the x-bar from left to right. A match condition will occur at each point except for the

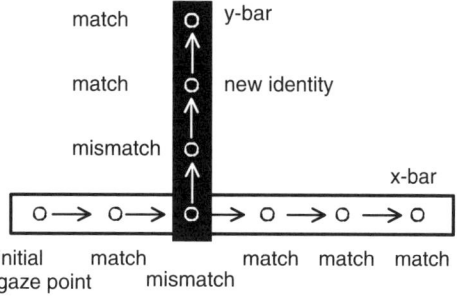

Figure 5.7 Predictive feedback generates identity via match and mismatch conditions

point where the black y-bar intersects the x-bar, as there the previous percept <white> does not match the present percept <black>. The perception/response loop is supposed to have a limited frequency response so that it functions like a lowpass filter. Therefore, if the black y-bar is traversed quickly then a new match condition will not have time to emerge. The old first-order prediction is retained and a match condition will be regained as soon as the white x-bar comes again into view. However, if the gaze begins to follow the black y-bar, then after another mismatch the prediction changes and match conditions will follow. What good is this? The match condition indicates that the same object is being seen or scanned; it provides an identity continuum for whatever object is being sensed.

This principle applies to moving objects as well. For instance a moving car should be recognized as the same even though its position changes. Those familiar with digital video processing know that this is a real problem for the computer. Special algorithms are required as this operation involves some kind of a comparison between the previous and present scenes. The moving object has to be detected and recognized over and over again, and it does not help that in the meantime the poor computer would surely also have other things to do.

The short-term memory loop function of the perception/response loop executes this kind of process in a direct and uncomplicated way. If the object moves, the object in the new position will be recognized as the same as the object in the previous position as it generates the match condition with the sustained memory of the previous percept, that is with the feedback representation. The movement may change the object's appearance a little. However, due to the nature of the feature signal representation all features will not change at the same time and thus enough match signals may be generated for the preservation of the identity. As the short-term memory is constantly updated, the object may eventually change completely, yet the identity will be preserved. In this way the perception/response loop is able to establish an identity to all perceived objects and this identity allows, for example, the visual tracking of an object; the object is successfully tracked when the match condition is preserved.

This process would also lead to the perception of the motion of objects when there is no real motion, but a series of snapshots, such as in a movie. Unfortunately artefacts such as the 'blinking lamps effect' would also appear. When two closely positioned lamps blink alternately the light seems to jump from one lamp to the other and the other way around, even though no real motion is involved. The identities of the lamps are taken as the same as they cause the perception of the same visual feature signals.

The short-term memory loop function of the perception/response loop also facilitates the comparison of perceived patterns. In Figure 5.8 the patterns B and C are to be compared to the pattern A. In this example the pattern A should be perceived first and the perception/response loop should sustain its feature signals at the feedback loop. Thereafter one of the patterns B and C should be perceived. The mismatch condition will occur when the pattern B is perceived after A and the match condition will occur when the pattern C is perceived after A. Thus the system has passed its first IQ test.

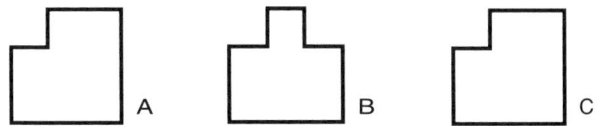

Figure 5.8 Comparison: which one of the patterns B and C is similar to A?

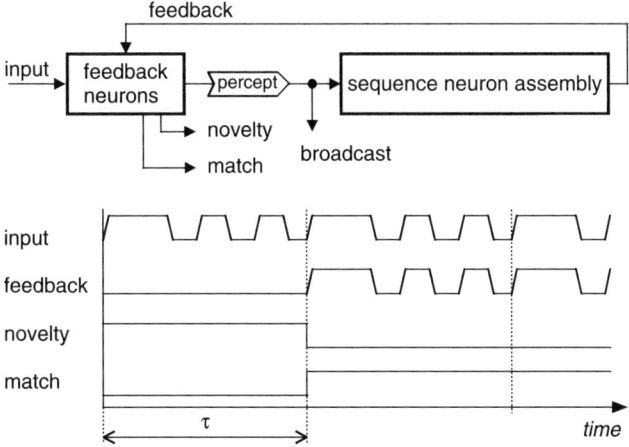

Figure 5.9 Higher-order prediction

The first-order prediction process is subject to the 'change blindness' phenomenon. Assume that certain patterns A, B and C are presented sequentially to the system. The pattern C is almost the same as the pattern A, but the pattern B is completely different, say totally black. The system should now detect the difference between the patterns A and B. This detection will fail if the pattern B persists long enough, so that the A pattern based prediction will fade away. Thereafter the pattern B will become as the next prediction and only the differences between B and C will cause mismatch conditions.

Nevertheless, the first-order prediction process is a very useful property of the perception/response loop. Higher-order prediction is possible if a sequence memory is inserted into the loop. This would allow the prediction of sequences such as melodies and rhythm.

In Figure 5.9 a sequence neuron assembly is inserted into the perception/response loop. This assembly will learn the incoming periodic sequence and will begin to predict it at a certain point. Initially the input sequence generates the novelty signal, but during successful prediction this novelty condition turns into the match condition.

5.5 KINESTHETIC PERCEPTION

Kinesthetic perception (kinesthesia) gives information about the position, motion and tension of body parts and joints in relation to each other. Proprioception is understood

82 MACHINE PERCEPTION

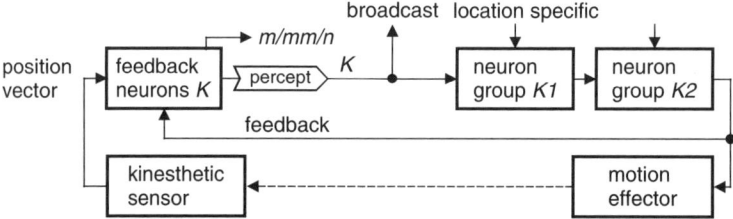

Figure 5.10 Kinesthetic perception in a motion control system

here as kinesthetic perception and balance perception together. Kinesthetic perception is related to motion and especially to motion control feedback systems.

In Figure 5.10 a typical motion control system is depicted. The kinesthetic posture is sensed by suitable sensors and the corresponding percept K is broadcast to the rest of the system. The neuron group $K2$ is a short-term memory and the neuron group $K1$ is a long-term memory. A location-specific representation may evoke a kinesthetic position vector, which is fed back to the feedback neurons and may thus become an 'imagined' kinesthetic position percept. This feedback would also correspond to the expected new kinesthetic percept, which would be caused by the system's subsequent action. The match/mismatch/novelty ($m/mm/n$) signals indicate the relationship between the feedback vector and the actually sensed kinesthetic position vector. The motion effector may also translate the evoked kinesthetic vector into the corresponding real mechanical position, which in turn would be sensed by the kinesthetic sensor.

In robotic applications various devices can be used as kinesthetic sensors to determine relative mechanical positions. The commonly used potentiometer provides this information as a continuous voltage value. In the motor control examples to follow the potentiometer is used, as its operation is easy to understand.

Kinesthetic perception may also be used for force or weight sensing, for instance for lifting actions. In that case a tension sensor would be used as the kinesthetic sensor and the motion effector would be commanded to provide a force that would cause the desired tension. In this case the internal feedback would represent the expected tension value. The match/mismatch signals would indicate that the actual and expected tensions match or do not match. The latter case would correspond to the 'empty milk carton effect', the unexpected lightness of a container that was supposed to be heavier. For smooth actions it is useful to command the execution force in addition to the motion direction.

Kinesthetic perception is also related to other sensory modalities. For vision, kinesthesia provides gaze direction information, which can also be used as 'a memory location address' for visually perceived objects.

5.6 HAPTIC PERCEPTION

The haptic or touch sense gives information about the world via physical contacts. In humans haptic sensors are embedded in the skin and are sensitive to pressure and vibration. Groups of haptic sensors can give information about the hardness, softness, surface roughness and texture of the sensed objects. Haptic sensors also allow shape sensing, the sensing of the motion of a touching object ('crawling bug') and the creation of a 'bodily self-image' (body knowledge).

Haptic shape sensing involves the combination of haptic and kinesthetic information and the short-term memorization of haptic percepts corresponding to each kinesthetic position. Shape sensing is not a passive reception of information, instead it is an active process of exploration as the sensing element (for instance a finger) must go through a series of kinesthetic positions and the corresponding haptic percepts must be associated with each position. If the surface of an object is sensed in this way then the sequence of the kinesthetic positions will correspond to the contours of that object. Haptic percepts of contours may also be associated with visual shape percepts and vice versa. The connection between the haptic and kinesthetic perception is depicted in Figure 5.11.

In Figure 5.11 the haptic input vector originates from a group of haptic sensors. The kinesthetic position vector represents the instantaneous relative position of the sensing part, for instance a finger. The neuron groups $H2$ and $K2$ are short-term memories that by virtue of their associative cross-connection sustain the recent haptic percept/position record.

The neuron groups $H1$ and $K1$ are long-term memories. A location-specific associative input at the $K1$ neuron group may evoke a kinesthetic position percept K and the motion effector may execute the necessary motion in order to reach that position. The match/mismatch/novelty ($m/mm/n$) output at the feedback neurons K would indicate the correspondence between the 'imagined' and actual positions.

The object-specific associative input at the $H1$ neuron group may evoke an expectation for the haptic features of the designated object. For instance, an object 'cat' might evoke percepts of <soft> and an object 'stone' might evoke percepts of <hard>. The match/mismatch/novelty ($m/mm/n$) output at the feedback neurons H would indicate the correspondence between the expected and actual haptic percepts.

An object ('a bug') that moves on the skin of the robot activates sequentially a number of touch sensors. This can be interpreted as motion if the outputs of the touch sensors are connected to motion detection sensors that can detect the direction of the change.

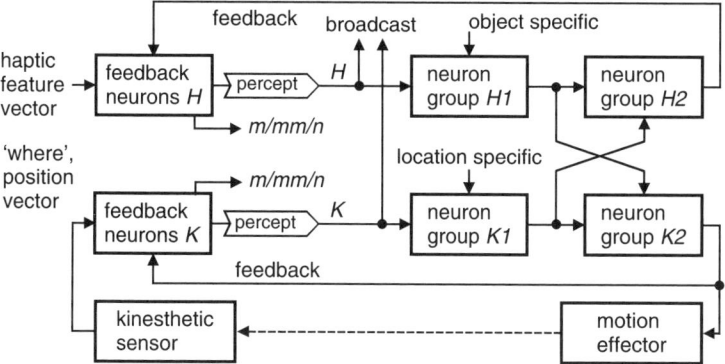

Figure 5.11 The connection between haptic and kinesthetic perception

A robot with a hand may touch various parts of its own body. If the 'skin' of the robot body is provided with haptic sensors then this touching will generate two separate haptic signals, one from the touching part, the finger, and one from the touched part. Thus the touched part will be recognized as a part belonging to the robot itself. Moreover, the kinesthetic information about the hand position in relation to the body may be associated with the haptic signal from the touched part of the body. Thus, later on, when something touches that special part of the body, the resulting haptic signal can evoke the associated kinesthetic information and consequently the robot will immediately be able to touch that part of the body. Via this kind of self-probing the robot may acquire a kinesthetic map, 'a body self-image', of its reachable body parts.

5.7 VISUAL PERCEPTION

5.7.1 Seeing the world out there

The eye and the digital camera project an image on to a photosensitive matrix, namely the retina and the array of photosensor elements. This is the actual visual image that is sensed. However, humans do not see and perceive things in that way. Instead of seeing an image on the retina humans perceive objects that are out there at various distances away while the retina and the related neural processing remain hidden and transparent. For a digital camera, even when connected to a powerful computer, this illusion has so far remained elusive. However, this is one of the effects that would be necessary for truly cognitive machines and conscious robots. How could this effect be achieved via artificial means and what would it take?

For a computer a digital image is a data file, and therefore is really nowhere. Its position is not inherently fixed to the photosensor matrix of the imaging device or to the outside world. In fact the computer does not even see the data as an image; it is just a file of binary numbers, available whenever requested by the program. Traditional computer vision extracts visual information by digital pattern recognition algorithms. It is also possible to measure the direction and distance of the recognized object, not necessarily by the image data only but by additional equipment such as ultrasonic range finders. Thereafter this numeric information could be used to compute trajectories for motor actions, like the grasping of an object. However, it is obvious that these processes do not really make the system see the world in the way that humans do, to be out there.

Here visual perception processes that inherently place the world out there are sought. Humans do not see images of objects, they believe to see the objects as they are. *Likewise the robot should not treat the visual information as coming from images of objects but as from the objects themselves; the process of imaging should be transparent.* Visual perception alone may not be able to place the objects out there, but a suitable connection to haptic and kinesthetic perception should provide the additional information. The combined effect should cause a visually perceived object to appear as one that can be reached out for and touched; it cannot be touched by touching the camera that gives the image. Also, the perceived shape and size of an object must conform to the shape and size that would be perceived via haptic perception.

The exact recognition of objects is secondary; initially it suffices that the robot sees that there are patterns and gestalts, 'things' out there. Thereafter it suffices that these 'things' remind them of something and seamlessly evoke possibilities for action. However, this is actually beyond basic perception and belongs to the next level of cognition.

In the following the visual perception process is outlined with the help of simple practical examples of the required processing steps.

5.7.2 Visual preprocessing

The purpose of visual preprocessing is to create visual feature signal vectors with meanings that are grounded to external world properties. A visual sensor with built-in neural style processing would be ideal for cognitive robots. As these are not readily available the use of conventional digital cameras is considered.

A digital camera generates pixel map images of the sensed environment. A pixel map is a two-dimensional array of picture elements (pixels) where each pixel is assigned with a number value that is proportional to the intensity of illumination of that point in the image.

Figure 5.12 depicts an $m \times n$ pixel map where each pixel depiction $P(i, j)$ describes the intensity of illumination at its position. Colour images have three separate pixel maps, one for each primary colour (red, green, blue; R, G, B) or alternatively one luminance (Y) component map and two colour difference (U, V) maps. In the following RGB maps are assumed when colour is processed, at other times the Y pixel map is assumed.

The task of visual perception is complicated by the fact that the image of any given object varies in its apparent size and shape and also the illumination may change.

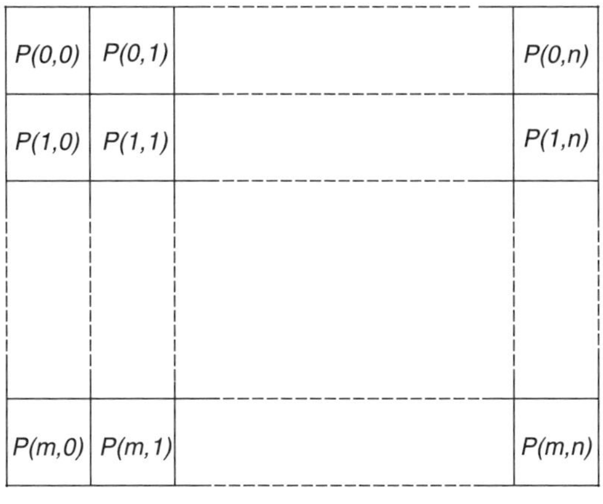

Figure 5.12 The pixel map

86 MACHINE PERCEPTION

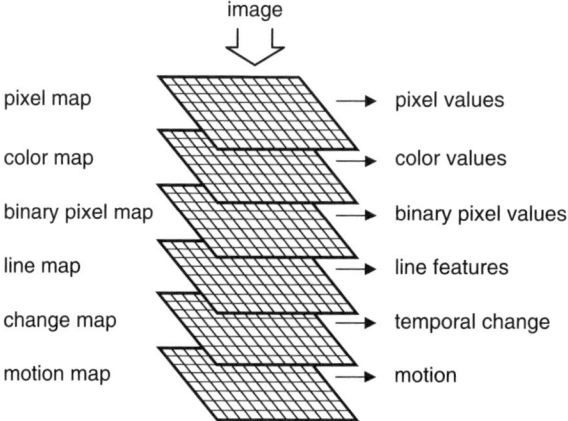

Figure 5.13 Visual feature maps

The pixel intensity map does not provide the requested information directly, and, moreover, is not generally compatible with the neuron group architecture. Therefore the pixel intensity map information must be dissected into maps that represent the presence or absence of the given property at each pixel position. Figure 5.13 depicts one possible set of visual feature maps.

Useful visual features could include colour values, binary pixel values, elementary lines, temporal change and spatial motion.

5.7.3 Visual attention and gaze direction

The information content of the visually sensed environment can be very high and consequently would require enormous processing capacity. In associative neural networks this would translate into very large numbers of neurons, synapses and interconnections. In the human eye and brain the problem is alleviated by the fact that only a very small centre area of the retina, the fovea, has high resolution while at the peripheral area the resolution is graded towards a very low value. This arrangement leads to a well-defined gaze direction and visual attention; objects that are to be accurately inspected visually must project on to the fovea. As a consequence the space of all possible gaze directions defines a coordinate system for the positions of the visually seen objects.

Humans believe that they perceive all their surroundings with the fullest resolution all of the time. In reality this is only an illusion. Actually only a very small area can be seen accurately at a time; the full-resolution illusion arises from the scanning of the environment by changing the gaze direction. Wherever humans turn their gaze they see everything with the full resolution. In this way the environment itself is used as a high-resolution visual memory.

The fovea arrangement can be readily utilized in robotic vision. The full-resolution pixel map may be subsampled into a new one with a high-resolution centre area and lower-resolution peripheral area, as shown in Figure 5.14. The low-resolution

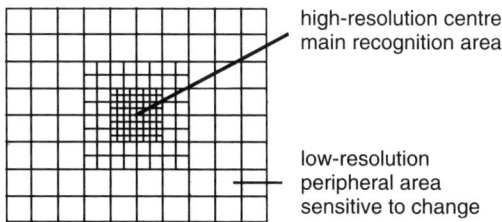

Figure 5.14 The division of the image area into a high-resolution centre area and a low-resolution peripheral area

peripheral area should be made sensitive to change and motion, which should be done in a way that would allow automatic gaze redirection to bring the detected change on to the fovea.

The high-resolution centre is the main area for object inspection and also defines the focus of visual attention. Gaze is directed towards the object to be inspected and consequently the image of the object is projected on to the high-resolution centre. This act now defines the relative positions of the parts of the object; the upper right and left part, the lower right and left part, etc. This will simplify the subsequent recognition process. For instance, when seeing a face, the relative position for the eyes, nose and mouth are now resolved automatically.

Objects are not only inspected statically; the gaze may seek and explore details and follow the contours of the object. In this way a part of the object recognition task may be transferred to the kinesthetic domain; different shapes and contours lead to different sequences of gaze direction patterns.

5.7.4 Gaze direction and visual memory

Gaze direction is defined here as the direction of the light ray that is projected on to the centre of the high-resolution area of the visual sensor matrix (fovea) and accordingly on to the focus of the primary visual attention. In a visual sensor like a video camera the gaze direction is thus along the optical axis. It is assumed that the camera can be turned horizontally and vertically (pan and tilt) and in this way the gaze direction can be made to scan the environment. Pan and tilt values are sensed by suitable sensors and these give the gaze direction relative to the rest of the mechanical body.

Gaze direction information is derived from the kinesthetic sensors that measure the eye (camera) direction. All possible gaze directions form coordinates for the seen objects. Gaze direction provides the 'where' information while the pixel perception process gives the 'what' information. The 'what' and 'where' percepts should be associated with each other continuously.

In Figure 5.15 the percepts of the visual features of an object constitute the 'what' information at the broadcast point V. The gaze direction percept that constitutes the 'where' information appears at the broadcast point Gd. These percepts are also broadcast to the rest of the system.

88 MACHINE PERCEPTION

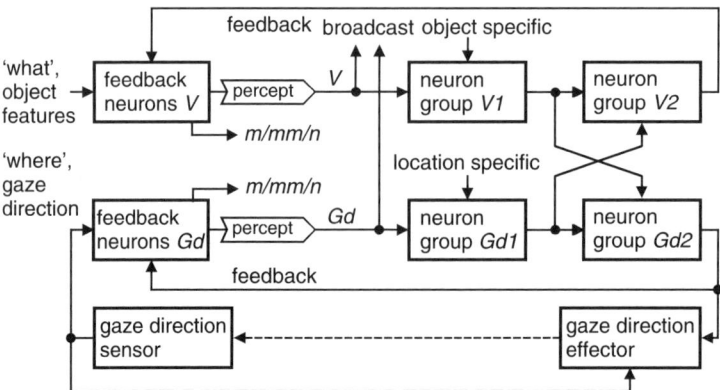

Figure 5.15 The association of visually perceived objects with a corresponding gaze direction

'What' and 'where' are associated with each other via the cross-connections between the neuron groups $Gd2$ and $V2$. The activation of a given 'what' evokes the corresponding location and vice versa. The location for a given object may change; therefore the neuron groups $V2$ and $Gd2$ must not create permanent associations, as the associations must be erased and updated whenever the location information changes.

Auxiliary representations may be associated with 'what' at the neuron group $V1$. This neuron group acts as a long-term memory, as the 'object-specific' representations (for instance a name of an object) correspond permanently to the respective visual feature vectors (caused by the corresponding object). Thus the name of an object would evoke the shape and colour features of that object. These in turn would be broadcast to the gaze direction neuron group $Gd2$ and if the intended object had been seen within the visual environment previously, its gaze direction values would be evoked and the gaze would be turned towards the object.

Likewise a 'location-specific' representation can be associated with the gaze direction vectors at the neuron group $Gd1$. These location-specific representations would correspond to the locations (up, down, to the left, etc.) and these associations should also be permanent. A location-specific representation would evoke the corresponding gaze direction values and the gaze would be turned towards that direction.

Gaze direction may be used as a mechanism for short-term or working memory. Imagined entities may be associated with imagined locations, that is with different gaze directions, and may thus be recalled by changing the gaze direction. An imagined location-specific vector will evoke the corresponding gaze direction. Normally this would be translated into the actual gaze direction by the gaze direction effector and the gaze direction sensor would give the percept of the gaze direction, which in turn would evoke the associated object at that direction at the neuron group $V2$. However, in imagination the actual motor acts that execute this gaze direction change are not necessary; the feedback to the gaze direction feedback neurons Gd is already able to evoke the imagined gaze direction percept. The generated mismatch signal at the gaze direction feedback neurons Gd will indicate that the direction percept does not correspond to the real gaze direction.

5.7.5 Object recognition

Signal vector representations represent objects as collections of properties or elementary features whose presence or absence are indicated by *one* and *zero*. A complex object has different features and properties at different positions; this fact must be included in the representation. The feature/position information is present in the feature maps and can thus be utilized in the subsequent recognition processes. Figure 5.16 gives a simplified example of this kind of representation.

In the framework of associative processing the recognition of an object does not involve explicit pattern matching; instead it relates to the association of another signal vector with it. This signal vector may represent a name, label, act or some other entity and the evocation of this signal vector would indicate that the presence of the object has been detected. This task can be easily executed by the associative neuron group, as shown in Figure 5.17.

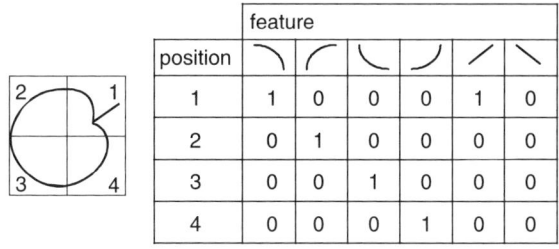

Figure 5.16 The representation of an object by its feature/position information

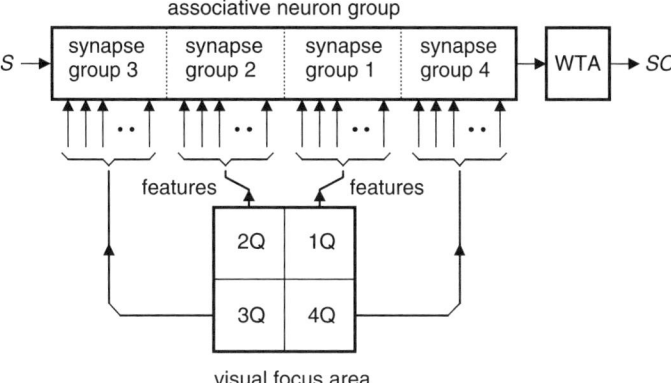

Figure 5.17 Encoding feature position information into object recognition

90 MACHINE PERCEPTION

In Figure 5.17 the visual focus area (fovea) is divided into four subareas 1Q, 2Q, 3Q, 4Q. Visual features are detected individually within these areas and are forwarded to an associative neuron group. This neuron group has specific synapse groups that correspond to the four subareas of the visual focus area. A label signal s may be associated with a given set of feature signals. For instance, if something like 'eyes' were found in the subareas 2Q and 1Q and something like a part of a mouth in the subareas 3Q and 4Q then a label signal s depicting a 'face' could be given. This example should illustrate the point that this arrangement encodes intrinsically the relative position information of the detected features. Thus, not only the detected features themselves but also their relative positions would contribute to the evocation of the associated label. In hardware implementations the actual physical wiring does not matter; the feature lines from each subarea do not have to go to adjacent synapses at the target neuron. Computer simulations, however, are very much simplified if discrete synaptic groups are used.

The above feature-based object recognition is not sufficient in general cases. Therefore it must be augmented by the use of feedback and inner models (gestalts). The role of gestalts in human vision is apparent in the illusory contours effect, such as shown in Figure 5.18.

Figure 5.18 shows an illusory white square on top of four black circles. The contours of the sides of this square appear to be continuous even though obviously they are not. The effect should vanish locally if one of the circles is covered. The perception of the illusory contours does not arise from the drawing because there is nothing in the empty places between the circles that could be taken to be a contour. Thus the illusory contours must arise from inner models.

The perception/response loop easily allows the use of inner models. This process is depicted in Figure 5.19, where the raw percept evokes one or more inner models. These models may be quite simple, consisting of some lines only, or they may be more complex, depicting for instance faces of other entities. The inner model signals are evoked at the neuron group $V1$ and are fed back to the feedback neuron group. The model signals will amplify the corresponding percept signals and will appear alone weakly where no percept signals exist. The match condition will be generated if there is an overall match between the sensory signals and the inner model. Sometimes there may be two or more different inner models that match the

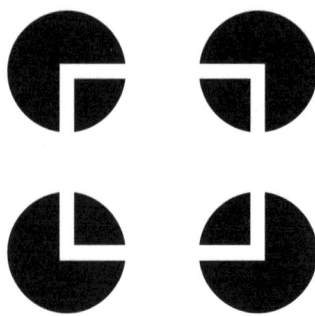

Figure 5.18 The illusory contours effect

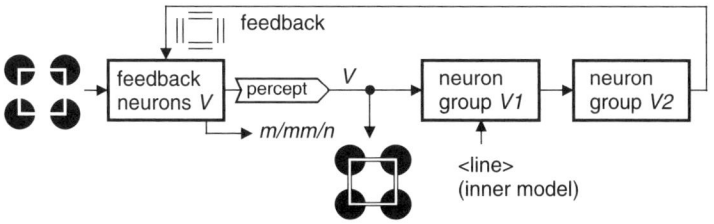

Figure 5.19 The use of inner models in visual perception

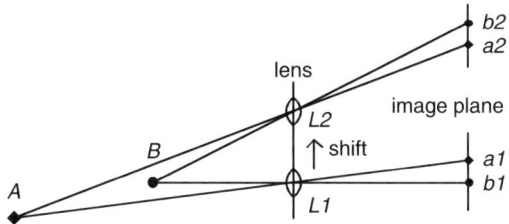

Figure 5.20 Camera movement shifts the lens and causes apparent motion

sensory signals. In that case the actual pattern percept may alternate between the models (the Necker cube effect). The inner model may be evoked by the sensory signals themselves or by context or expectation.

The segregation of objects from a static image is difficult, even with inner models. The situation can be improved by exploratory actions such as camera movement. Camera movement shifts the lens position, which in turn affects the projection so that objects at different distances seem to move in relation to each other (Figure 5.20).

In Figure 5.20 the lens shifts from position $L1$ to $L2$ due to a camera movement. It can be seen that the relative positions of the projected images of the objects A and B will change and B appears to move in front of A. This apparent motion helps to segregate individual objects and also gives cues about the relative position and distances of objects.

5.7.6 Object size estimation

In an imaging system such as the eye and camera the image size of an object varies on the image plane (retina) according to the distance of the object. The system must not, however, infer that the actual size of the object varies; instead the system must infer that the object has a constant size and the apparent size change is only due to the distance variation. Figure 5.21 shows the geometrical relationship between the image plane image size and the distance to the object.

According to the thin lens theory light rays that pass through the centre of the lens are not deflected. Thus

$$h1/f = H/d1 \tag{5.12}$$

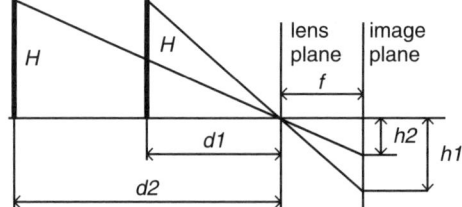

Figure 5.21 The effect of object distance on the image size at the image plane

where

$h1$ = image height at the image plane
f = focal length of the lens
H = actual object height
$d1$ = distance to the object

The image height at the image plane will be

$$h1 = H * f/d1 \qquad (5.13)$$

Thus, whenever the distance to the object doubles the image height halves. Accordingly the object height will be

$$H = h1 * d1/f \qquad (5.14)$$

The focal length can be considered to be constant. Thus the system may infer the actual size of the object from the image size if the distance can be estimated.

5.7.7 Object distance estimation

Object distance estimations are required by the motor systems, hands and motion, so that the system may move close enough to the visually perceived objects and reach out for them. Thus an outside location will be associated with the visually perceived objects, and the visual distance will also be associated with the motor action distances.

The distance of the object may be estimated by the image size at the image plane if the actual size of the object is known. Using the symbols of Figure 5.21 the object distance is

$$d1 = f * H/h1 \qquad (5.15)$$

Thus, the smaller the object appears the further away it is. However, usually more accurate estimations for the object distance are necessary.

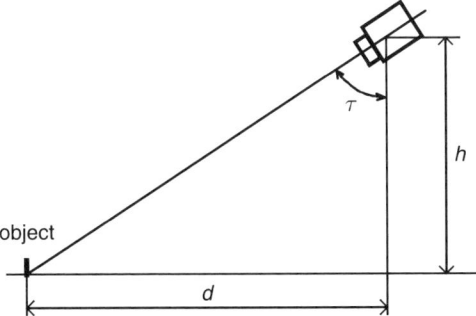

Figure 5.22 Distance estimation: a system must look down for objects near by

The gaze direction angle may be used to determine the object distance. In a simple application a robotic camera may be situated at the height h from the ground. The distance of the nearby objects on the ground may be estimated by the tilt angle τ of the camera (Figure 5.22). According to Figure 5.22 the distance to the object can be computed as follows:

$$d = h^* \tan \tau \tag{5.16}$$

where

$d =$ distance to the object
$h =$ height of the camera position
$\tau =$ camera tilt angle

Should the system do this computation? Not really. The camera tilt angle τ should be measured and this value could be used directly as a measure of the distance.

Binocular distance estimation is based on the use of two cameras that are placed a small distance apart from each other. The cameras are symmetrically turned so that both cameras are viewing the same object; in each camera the object in question is imaged by the centre part of the image sensor matrix (Figure 5.23).

The system shown in Figure 5.23 computes the difference between the high-resolution centre (fovea) images of the left and right camera. Each camera is turned symmetrically by a motor that is controlled by the difference value. The correct camera directions (convergence) are achieved when the difference goes to zero. If the convergence fails then the left image and right image will not spatially overlap and two images of the same object will be perceived by the subsequent circuitry. False convergence is also possible if the viewed scene consists of repeating patterns that can give zero difference at several α angle values.

According to Figure 5.23 the distance to the object can be computed as follows:

$$d = L^* \tan \alpha \tag{5.17}$$

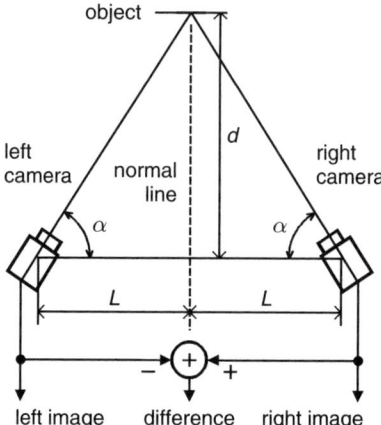

Figure 5.23 Binocular distance estimation

where

d = distance to the object
L = half of the distance between the left camera and the right camera
α = camera turn angle

Again, there is no need to compute the actual value for the distance d. The angle α can be measured by a potentiometer or the like and this value may be used associatively.

The use of two cameras will also provide additional distance information due to the stereoscopic effect; during binocular convergence only the centre parts of the camera pictures match and systematic mismatches occur elsewhere. These mismatches are related to the relative distances of the viewed objects.

5.7.8 Visual change detection

Visual change detection is required for focusing of visual attention and the detection of motion. In a static image the intensity value of each pixel remains the same regardless of its actual value. Motion in the sensed area of view will cause temporal change in the corresponding pixels. When an object moves from a position A to a position B, it disappears at the position A, allowing the so far hidden background to become visible. Likewise, the object will appear at the position B covering the background there. A simple change detector would indicate pixel value change regardless of the nature of the change, thus change would be indicated at the positions A and B. Usually the new position of the moving object is more interesting and therefore the disappearing and appearing positions should be differentiated.

When the object disappears the corresponding pixel values turn into the values of the background and when the object appears the corresponding pixel values turn into

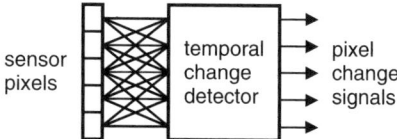

Figure 5.24 The temporal change detector

values that are different from the background. Thus two comparisons are required: the temporal comparison that detects the change of the pixel value and the spatial comparison that detects the pixel value change in relation to unchanged nearby pixels. Both of these cases may be represented by one signal per pixel. This signal has the value of zero if no change is detected, a high positive value if appearance is detected and a low positive value if disappearance is detected. The temporal change detector is depicted in Figure 5.24.

What happens if the camera turns? Obviously all pixel values may change as the projected image travels over the sensor. However, nothing appears or disappears and there is no pixel value change in relation to nearby pixels (except for the border pixels). Therefore the change detector should output zero-valued signals.

5.7.9 Motion detection

There are various theories about motion detection in human vision. Analogies from digital video processing would suggest that the motion of an object could be based on the recognition of the moving object at subsequent locations and the determination of the motion based on these locations. However, the continuous search and recognition of the moving object is computationally heavy and it may be suspected that the eye and the brain utilize some simpler processes. If this is so, then cognitive machines should also use these.

The illusion of apparent motion suggests that motion detection in human vision may indeed be based on rather simple principles, at least for some part. The illusion of apparent motion can be easily demonstrated by various test arrangements, for instance by the image pair of Figure 5.25. The images 1 and 2 of Figure 5.25 should be positioned on top of each other and viewed sequentially (this can be done, for example, by the Microsoft Powerpoint program). It can be seen that the black circle appears to move to the right and change into a square and vice versa. In fact the circle and the square may have different colours and the motion illusion still persists.

The experiment of Figure 5.25 seems to show that the motion illusion arises from the simultaneous disappearance and appearance of the figures and not so much from any pattern matching and recognition. Thus, motion detection would rely on temporal change detection. This makes sense; one purpose of motion detection is the possibility to direct gaze towards the new position of the moving object and this would be the purpose of visual change detection in the first place. Thus, the detected motion of an object should be associated with the corresponding motor command signals that would allow visual tracking of the moving object.

96 MACHINE PERCEPTION

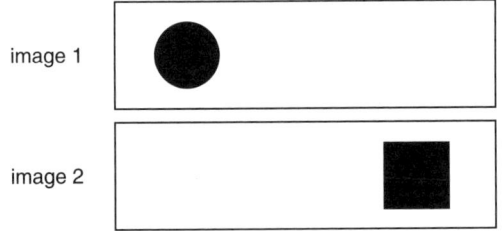

Figure 5.25 Images for the apparent motion illusion

The afterimage dot experiment illustrates further the connection between eye movement and the perceived motion of a visually perceived object. Look at a bright spot for a minute or so and then darken the room completely. The afterimage of the bright spot will be seen. Obviously this image is fixed on the retina and cannot move. However, when you move your eyes the dot seems to move and is, naturally, always seen in the direction of the gaze. The perceived motion cannot be based on any visual motion cue, as there is none; instead the motion percept corresponds to the movement of the eyes. If the eyes are anaesthetized so that they could not actually move, the dot would still be seen moving according to the tried eye movements (Goldstein, 2002, p. 281). This suggests that the perception of motion arises from the intended movement of the eyes.

The corollary discharge theory proposes a possible neural mechanism for motion detection (Goldstein, 2002 pp. 279–280). According to this theory the motion signal is derived from a retinal motion detector. The principle of the corollary discharge theory is depicted in Figure 5.26.

In Figure 5.26 *IMS* is the detected image movement signal, *MS* is the commanded motor signal and the equivalent *CDS* is the so-called corollary discharge signal that controls the motion signal towards the brain. The retinal motion signal *IMS* is inhibited by the *CDS* signal if the commanded motion of the eyeball causes the detected motion on the retina. In that case the visual motion would be an artefact created by the eye motion, as the projected image travels on the retina, causing the motion detector to output false motion *IMS*. In this simplified figure it is assumed that the retinal motion *IMS* and the corollary discharge *CDS* are binary and have one-to-one correspondence by means that are not considered here. In that case an exclusive-OR operation will execute the required motion signal inhibition.

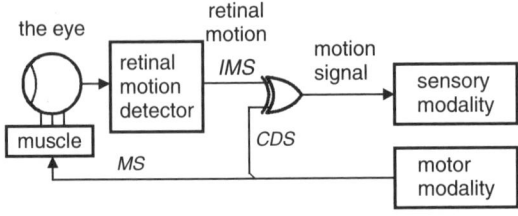

Figure 5.26 The corollary discharge model for motion detection

VISUAL PERCEPTION

According to the corollary discharge theory and Figure 5.26 motion should be perceived also if the eyeball moves without the motor command (due to external forces, no *CDS*) and also if the eyeball does not move (due to anaesthetics, etc.) even though the motor command is sent (*IMS* not present, *CDS* present). Experiments seem to show that this is the case.

The proposed artificial architecture that captures these properties of the human visual motion perception is presented in Figure 5.27. According to this approach the perceived visual motion is related to the corresponding gaze direction motor commands. After all, the main purpose of motion detection is facilitation of the tracking of the moving object by gaze and consequently the facilitation of grabbing actions if the moving object is close enough.

In Figure 5.27 the image of a moving object travels across the sensory pixel array. This causes the corresponding travelling temporal change on the sensor pixel array. This change is detected by the temporal change detector of Figure 5.24. The output of the temporal change detector is in the form of an active signal for each changed pixel and as such is not directly suitable for gaze direction control. Therefore an additional circuit is needed, one that transforms the changed pixel information into absolute direction information. This information is represented by single signal vectors that indicate the direction of the visual change in relation to the system's straight-ahead direction. Operation of this circuit is detailed in Chapter 6 'Motor Actions for Robots' in Section 6.4, 'gaze direction control'.

Gaze direction is controlled by the gaze direction perception/response feedback loop. In Figure 5.27 this loop consists of the feedback neuron group Gd, the neuron groups $Gd1a$ and $Gd1b$ and the Winner-Takes-All (WTA circuit). The loop senses the gaze direction (camera direction) by a suitable gaze direction sensor. The gaze direction effector consisting of pan and tilt motor systems turns the camera

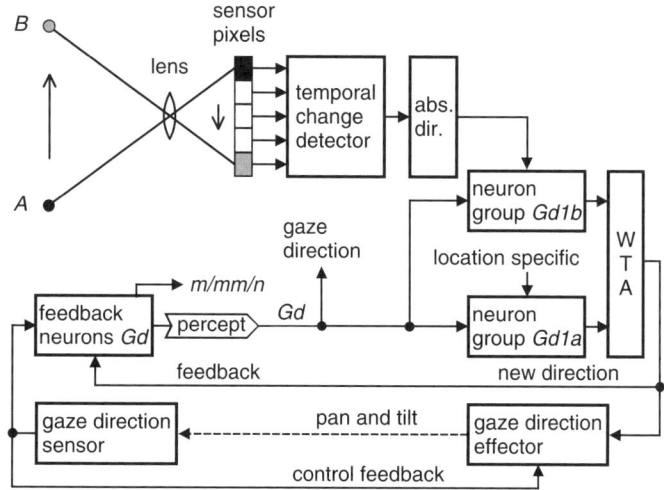

Figure 5.27 Motion detection as gaze direction control

towards the direction that is provided by the gaze direction perception/response feedback loop.

The absolute directions are associated with gaze direction percepts at the neuron group $Gd1b$. Thus a detected direction of a visual change is able to evoke the corresponding gaze direction. This is forwarded to the gaze direction effector, which now is able to turn the camera towards that direction and, if the visual change direction moves, is also able to follow the moving direction. On the other hand, the desired gaze direction is also forwarded to the feedback neurons Gd. These will translate the desired gaze direction into the gaze direction percept GdP. A location-specific command at the associative input of the neuron group $GD1a$ will turn the camera and the gaze towards commanded directions (left, right, up, down, etc.).

Whenever visual change is detected, the changing gaze direction percept GdP reflects the visual motion of the corresponding object. Also when the camera turning motor system is disabled, the gaze direction percept GdP continues to reflect the motion by virtue of the feedback signal in the gaze direction perception/response feedback loop.

False motion percepts are not generated during camera pan and tilt operations because the temporal change detector does not produce output if only the camera is moving. Advanced motion detection would involve the detection and prediction of motion patterns and the use of motion models.

5.8 AUDITORY PERCEPTION

5.8.1 Perceiving auditory scenes

The sound that enters human ears is the sum of the air pressure variations caused by all nearby sound sources. These air pressure variations cause the eardrums to vibrate and this vibration is what is actually sensed and transformed into neural signals. Consequently, the point of origin for the sound sensation should be the ear and the sound itself should be a cacophony of all contributing sounds. However, humans do not perceive things to be like that, instead they perceive separate sounds that come from different outside directions; an auditory scene is perceived in an apparently effortless and direct way. In information processing technology the situation is different. Various sound detection, recognition and separation methods exist, but artificial auditory scene analysis has remained notoriously difficult. Consequently, the lucid perception of an auditory scene as separate sounds out there has not yet been replicated in any machine. Various issues of auditory scene analysis are presented in depth, for instance, in Bregman (1994), but the lucid perception of auditory scenes is again one of the effects that would be necessary for cognitive and conscious machines.

The basic functions of the artificial auditory perception process would be:

1. Perceive separate sounds.

2. Detect the arrival directions of the sounds;

3. Estimate the sound source distance.

4. Estimate the sound source motion.

5. Provide short-term auditory (echoic) memory.

6. Provide the perception of the sound source locations out there.

However, it may not be necessary to emulate the human auditory perception process completely with all its fine features. A robot can manage with less.

5.8.2 The perception of separate sounds

The auditory scene usually contains a number of simultaneous sounds. The purpose of auditory preprocessing is the extraction of suitable sound features that allow the separation of different sounds and the focusing of auditory attention on any of these. The output from the auditory preprocesses should be in the form of signal vectors with meanings that are grounded in auditory properties.

It is proposed here that a group of frequencies will appear as a separate single sound if attention can be focused on it and, on the other hand, it cannot be resolved into its frequency components if attention cannot be focused separately on the individual auditory features. Thus *the sensation of a single sound would arise from single attention*.

A microphone transforms the incoming sum of sounds into a time-varying voltage, which can be observed by an oscilloscope. The oscilloscope trace (waveform) shows the intensity variation over a length of time allowing the observation of the temporal patterns of the sound. Unfortunately this signal does not easily allow the focusing of attention on the separate sounds, thus it is not really suited for sound separation.

The incoming sum of sounds can also be represented by its frequency spectrum. The spectrum of a continuous periodic signal consists of a fundamental frequency sine wave signal and a number of harmonic frequency sine wave signals. The spectrum of a transient sound is continuous. The spectrum analysis of sound would give a large number of frequency component signals that can be used as feature signals.

Spectrum analysis can be performed by Fourier transform or by a bandpass filter or resonator banks, which is more like the way of the human ear. If the auditory system were to resolve the musical scale then narrow bandpass filters with less than 10 % bandwidth would be required. This amounts to less than 5 Hz bandwidth at 50 Hz centre frequency and less than 100 Hz bandwidth at 1 kHz centre frequency. However, the resolution of the human ear is much better than that, around 0.2 % for frequencies below 4000 Hz (Moore, 1973).

The application of the perception/response feedback loop to artificial auditory perception is outlined in the following. It is first assumed that audio spectrum analysis is executed by a bank of narrow bandpass filters (or resonators) and the output from each filter is rectified and lowpass filtered. These bandpass filters should cover the whole frequency range (the human auditory frequency range is nominally

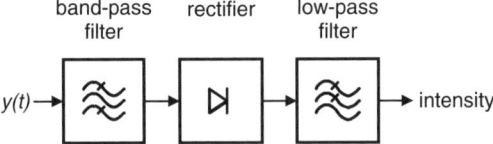

Figure 5.28 The detection of the intensity of a narrow band of frequencies

20–20 000 Hz; for many applications 50–5000 Hz or even less should be sufficient). Bandpass filtering should give a positive signal that is proportional to the intensity of the narrow band of frequencies that has passed the bandpass filter (Figure 5.28).

The complete filter bank contains a large number of the circuits shown in Figure 5.28. The output signals of the filter bank are taken as the feature signals for the perception/response feedback loop. The intensities of the filter bank output signals should reflect the intensities of the incoming frequencies.

In Figure 5.29 the filter bank has the outputs $f_0, f_1, f_2, \ldots, f_n$ that correspond to the frequencies of the bandpass filters. The received mix of sounds manifests itself as a number of active f_i signals. These signals are forwarded to the feedback neuron group, which also receives feedback from the inner neuron groups as the associative input. The perceived frequency signals are $pf_0, pf_1, pf_2, \ldots, pf_n$. The intensity of these signals depends on the intensities of the incoming frequencies and on the effect of the feedback signals.

Each separate sound is an auditory object with components, auditory features that appear simultaneously. Just like in the visual scene, many auditory objects may exist simultaneously and may overlap each other. As stated before, a single sound is an auditory object that can be attended to and selected individually. In the outlined system the function of attention is executed by thresholds and signal intensities. Signals with a higher intensity will pass thresholds and thus will be selected.

According to this principle it is obvious that the component frequency signals of any loud sound, even in the presence of background noise, will capture attention and the sound will be treated as a whole. Thus the percept frequency signals $pf_i, \ldots, pf_k$ of the sound would be associated with other signals in the system and possibly the other way round. In this way the components of the sound would be bound together.

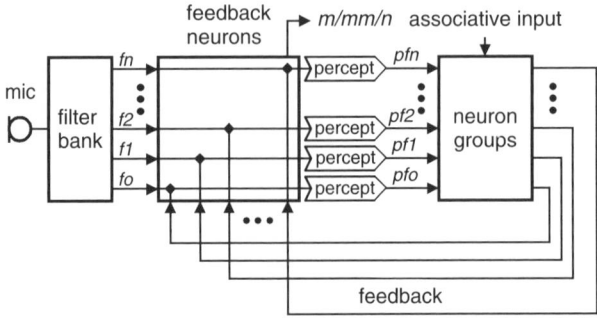

Figure 5.29 The perception of frequencies (simplified)

All sounds are not louder than the background or appear in total silence. New sounds especially should be able to capture attention, even if they are not louder than the background. For this purpose an additional circuit could be inserted in the output of the filter bank. This circuit would temporarily increase the intensity of all new signals.

In the perception/response feedback loop the percept signal intensities may also be elevated by associative input signals as described before. This allows the selection, prediction and primed expectation of sounds by the system itself.

5.8.3 Temporal sound pattern recognition

Temporal sound patterns are sequences of contiguous sounds that originate from the same sound source. Spoken words are one example of temporal sound patterns. The recognition of a sound pattern involves the association of another signal vector with it. This signal vector may represent a sensory percept or some other entity and the evocation of this signal vector would indicate that the presence of the object has been detected. A temporal sound pattern is treated here as a temporal sequence of sound feature vectors that represent the successive sounds of the sound pattern. The association of a sequence of vectors with a vector can be executed by the sequence-label circuit of Chapter 4.

In Figure 5.30 the temporal sound pattern is processed as a sequence of sound feature vectors. The registers start to capture sound feature vectors from the beginning of the sound pattern. The first sound vector is captured by the first register, the second sound vector by the next and so on until the end of the sound pattern. Obviously the number of available registers limits the length of the sound patterns that can be completely processed. During learning the captured sound vectors at their proper register locations are associated with a label vector S or a label signal s. During recognition the captured sound vectors evoke the vector or the signal that is most closely associated with the captured sound vectors.

Temporal sound pattern recognition can be enhanced by the detection of the sequence of the temporal intervals of the sound pattern (the rhythm), as described in Chapter 4. The sequence of these intervals could be associated with the label vector S or the signal s as well. In some cases the sequence of the temporal intervals alone might suffice for recognition of the temporal sound pattern.

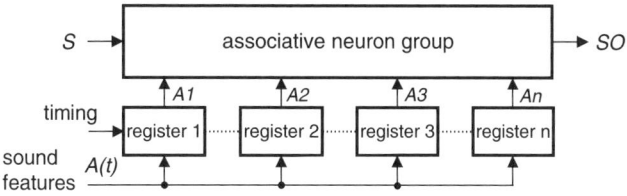

Figure 5.30 The recognition of a temporal sound pattern

5.8.4 Speech recognition

The two challenges of speech recognition are the recognition of the speaker and the recognition of the spoken words. Speech recognition may utilize the special properties of the human voice. The fundamental frequency or pitch of the male voice is around 80–200 Hz and the female pitch is around 150–350 Hz. Vowels have a practically constant spectrum over tens of milliseconds. The spectrum of a vowel consists of the fundamental frequency component (the pitch) and a large number of harmonic frequency components that are separated from each other by the pitch frequency. The intensity of the harmonic components is not constant; instead there are resonance peaks that are called formants. In Figure 5.31 the formants are marked as $F1$, $F2$ and $F3$.

The identification of a vowel can be aided by determination of the relative intensities of the formants. Determination of the relative formant intensities V_{F2}/V_{F1}, V_{F3}/V_{F1} etc., would seem to call for division operation (analog or digital), which is unfortunate as direct division circuits are not very desirable. Fortunately there are other possibilities; one relative intensity detection circuit that does not utilize division is depicted in Figure 5.32. The output of this circuit is in the form of a single signal representation.

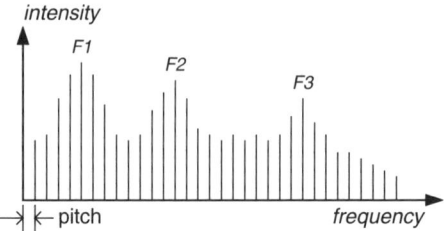

Figure 5.31 The spectrum of a vowel

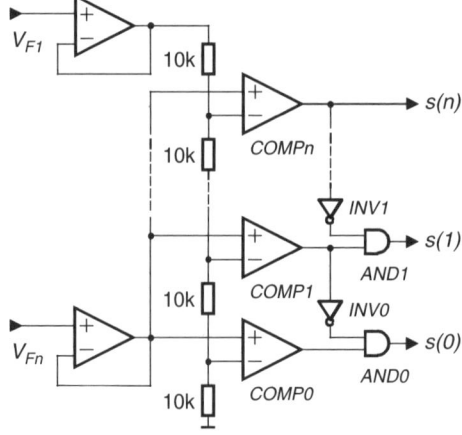

Figure 5.32 A circuit for the detection of the relative intensity of a formant

The circuit of Figure 5.32 determines the relative formant intensity V_{Fn} in relation to the formant intensity V_{F1}. The intensity of the V_{Fn} signal is compared to the fractions of the V_{F1} intensity. If the intensity of the formant Fn is very low then it may be able to turn on only the lowest comparator $COMP0$. If it is higher, then it may also be able to turn on the next comparator, $COMP1$, etc. However, here a single signal representation is desired. Therefore inverter/AND circuits are added to the output. They inhibit all outputs from the lower value comparators so that only the output from the highest value turned-on comparator may appear at the actual output. This circuit operates on relative intensities and the actual absolute intensity levels of the formants do not matter.

It is assumed here that the intensity of a higher formant is smaller that that of the lowest formant; this is usually the case. The circuit can, however, be easily modified to accept higher intensities for the higher formants.

The formant centre frequencies tend to remain the same when a person speaks with lower and higher pitches. The formant centre frequencies for a given vowel are lower for males and higher for females. The centre frequencies and intensities of the formants can be used as auditory speech features for phoneme recognition. The actual pitch may be used for speaker identification along with some other cues. The pitch change should also be detected. This can be used to detect question sentences and emotional states.

Speech recognition is notoriously difficult, especially under noisy conditions. Here speech recognition would be assisted by feedback from the system. This feedback would represent context- and situation model-generated expectations.

5.8.5 Sound direction perception

Sound direction perception is necessary for a perceptual process that places sound sources out there as the objects of the auditory scene. A straightforward method for sound direction perception would be the use of one unidirectional microphone and a spectrum analyser for each direction. The signals from each microphone would inherently contain the direction information; the sound direction would be the direction of the microphone. This approach has the benefit that no additional direction detection circuits would be required, as the detected sounds would be known to originate from the fixed direction of the corresponding microphone. Multiple auditory target detection would also be possible; sound sources with an identical spectrum but a different direction would be detected as separate targets. This approach also has drawbacks. Small highly unidirectional microphones are difficult to build. In addition, one audio spectrum analyser, for instance a filter bank, is required for each direction. The principle of sound direction perception with an array of unidirectional microphones is presented in Figure 5.33.

In Figure 5.33 each unidirectional microphone has its own filter bank and perception/response loop. This is a simplified diagram that gives only the general principle. In reality the auditory spectrum should be processed into further auditory features that indicate, for instance, the temporal change of the spectrum.

104 MACHINE PERCEPTION

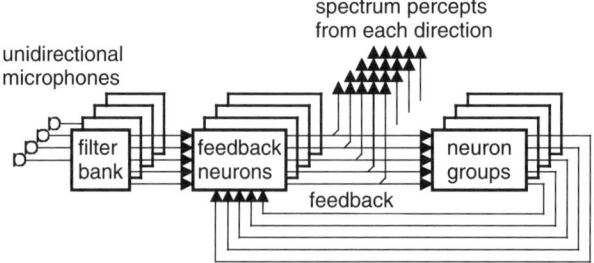

Figure 5.33 Sound direction perception with an array of unidirectional microphones

Sound direction can also be determined by two omnidirectional microphones that are separated from each other by an acoustically attenuating block such as the head. In this approach sound directions are synthesized by direction detectors. Each frequency band requires its own direction detector. If the sensed audio frequency range is divided into, say, 10 000 frequency bands, then 10 000 direction detectors are required. Nature has usually utilized this approach as ears and audio spectrum analysers (the cochlea) are rather large and material-wise expensive while direction detectors can be realized economically as subminiature neural circuits. This approach has the benefit of economy, as only two spectrum analysers are required. Unfortunately there is also a drawback; multiple auditory target detection is compromised. Sound sources with identical spectrum but different directions cannot be detected as separate targets, instead one source with a false sound direction is perceived. This applies also to humans with two ears. This shortcoming is utilized in stereo sound reproduction. The principle of the two-microphone sound direction synthesis is depicted in Figure 5.34.

In Figure 5.34 each filter bank outputs the intensities of the frequencies from the lowest frequency f_L to the highest frequency f_H. Each frequency has its own

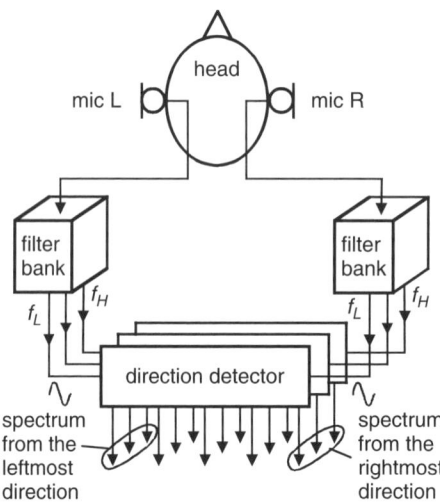

Figure 5.34 Sound direction synthesis with two microphones

direction detector. Here only three direction detectors are depicted; in practice a large number of direction detectors would be necessary. Each direction detector outputs only one signal at a time. This signal indicates the synthesized direction by its physical location. The intensity of this signal indicates the amplitude of the specific frequency of the sound. The signal array from the same direction positions of the direction detectors represents the spectrum of the sound from that direction. Each direction detector should have its own perception/response loop as depicted in Figure 5.35.

In Figure 5.35 each perception/response loop processes direction signals from one direction detector. Each direction detector processes only one frequency and if no tone with that frequency exists, then there is no output. Each signal line is hardwired for a specific arrival direction for the tone, while the intensity of the signal indicates the intensity of the tone. The direction detector may output only one signal at a time.

Within the perception/response loop there are neuron groups for associative priming and echoic memory. The echoic memory stores the most recent sound pattern for each direction. This sound pattern may be evoked by attentive selection of the direction of that sound. This also means that a recent sound from one direction cannot be reproduced as an echoic memory coming from another direction. This is useful, as obviously the cognitive entity must be able to remember where the different sounds came from and also when the sounds are no longer there.

In Figure 5.35 the associative priming allows (a) the direction information to be overridden by another direction detection system or visual cue and (b) visually assisted auditory perception (this may lead to the McGurk effect; see McGurk and MacDonald, 1976; Haikonen, 2003a, pp. 200–202).

Next the basic operational principles of possible sound direction detectors are considered.

5.8.6 Sound direction detectors

In binaural sound direction detection there are two main parameters that allow direction estimation, namely the intensity difference and the arrival time difference

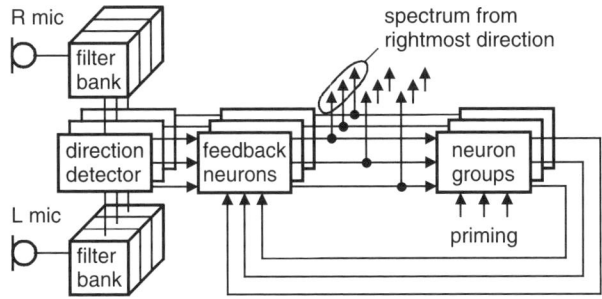

Figure 5.35 Sound perception with sound direction synthesis

of the sound between the ears. These effects are known as the 'interaural intensity difference' (IID) (sometimes also referred to as the amplitude or level difference) and the 'interaural time difference' (ITD). Additional sound direction cues can be derived from the direction sensitive filtering effects of the outer ears and the head generally. In humans these effects are weak while some animals do have efficient outer ears. In the following these additional effects are neglected and only sound direction estimation in a machine by the IID and ITD effects is considered. The geometry of the arrangement is shown in Figure 5.36.

In Figure 5.36 a head, real or artificial, is assumed with ears or microphones on each side. The distance between the ears or microphones is marked as L. The angle between the incoming sound direction and the head direction is marked as δ. If the angle δ is positive, the sound comes from the right; if it is negative, the sound comes from the left. If the angle δ is $0°$ then the sound comes from straight ahead and reaches both ears simultaneously. At other angle values the sound travels unequal distances, the distance difference being Δd, as indicated in Figure 5.36. This distance difference is an idealized approximation; the exact value would depend on the shape and geometry of the head. The arrival time difference is caused by the distance difference while the intensity difference is caused by the shading effect of the head. These effects on the reception of a sine wave sound are illustrated in Figure 5.37.

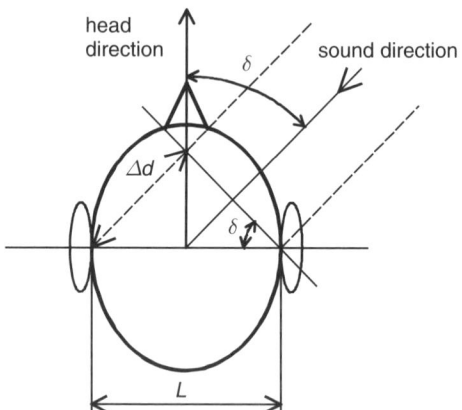

Figure 5.36 Binaural sound direction estimation

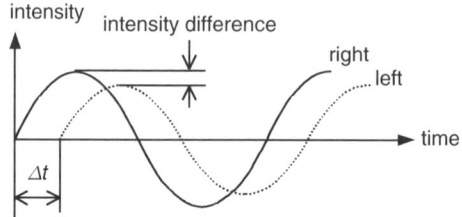

Figure 5.37 The effect of the head on the sound intensity and delay for a sine wave sound

The sound arrival direction angle can be computed by the arrival time difference using the markings of Figure 5.36 as follows:

$$\delta = arcsin(\Delta d/L) \tag{5.18}$$

Where

$\Delta d =$ distance difference for sound waves
$L =$ distance between ears (microphones)
$\delta =$ sound arrival direction angle

On the other hand,

$$\Delta d = \Delta t * v \tag{5.19}$$

Where

$\Delta t =$ delay time
$v =$ speed of sound $\approx (331.4 + 0.6*Tc)$m/s, where $Tc =$ temperature in Celsius

Thus

$$\delta = arcsin(\Delta t * v/L) \tag{5.20}$$

It can be seen that the computed sound direction is ambiguous as $sin(90° - x) = sin(90° + x)$. For instance, if $(\Delta t * v/L) = 0.5$ then δ may be 30° or 150° ($x = 60°$) and, consequently, the sound source may be in front of or behind the head. Likewise, if $\Delta t = 0$ then $\delta = arcsin(0) = 0°$ or 180°.

Another source of ambiguity sets in if the delay time is longer than the period of the sound. The maximum delay occurs when the sound direction angle is 90° or $-90°$. In those cases the distance difference equals the distance between the ears ($\Delta d = L$) and from Equation (5.19) the corresponding time difference is found as

$$\Delta t = \Delta d/v = L/v$$

If $L = 22$ cm then the maximum delay is about $0.22/340$ s $= 0.65$ ms, which corresponds to the frequency of 1540 Hz. At this frequency the phases of the direct and delayed sine wave signals will coincide and consequently the time difference will be falsely taken as zero. Thus it can be seen that the arrival time difference method is not suitable for continuous sounds of higher frequencies.

The artificial neural system here accepts only signal vector representations. Therefore the sound direction should be represented by a single signal vector that has one signal for each discrete direction. There is no need to actually compute the sound direction angle δ and transform this into a signal vector representation as

108 MACHINE PERCEPTION

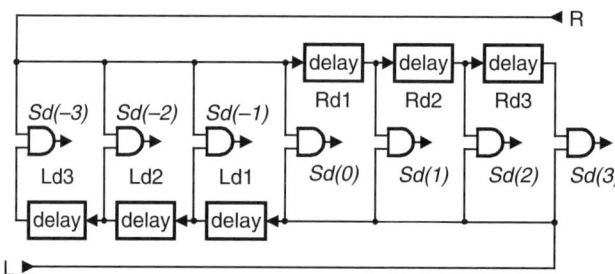

Figure 5.38 An arrival time comparison circuit for sound direction estimation

the representation can be derived directly from a simple arrival time comparison. The principle of a circuit that performs this kind of sound direction estimation is depicted in Figure 5.38. In this circuit the direction is represented by the single signal vector $\{Sd(-3), Sd(-2), Sd(-1), Sd(0), Sd(1), Sd(2), Sd(3)\}$. Here $Sd(0) = 1$ would indicate the direction $\delta = 0°$ and $Sd(3) = 1$ and $Sd(-3) = 1$ would indicate directions to the right and left respectively. In practice there should be a large number of delay lines with a short delay time.

In the circuit of Figure 5.38 it is assumed that a short pulse is generated at each leading edge of the right (R) and the left (L) input signals. The width of this pulse determines the minimum time resolution for the arrival delay time. If the sound direction angle is 0° then there will be no arrival time difference and the right and left input pulses coincide directly. This will be detected by the AND circuit in the middle and consequently a train of pulses will appear at the $Sd(0)$ output while other outputs stay at the zero level. If the sound source is located to the right of the centreline then the distance to the left microphone is longer and the left signal is delayed by a corresponding amount. The delay lines Rd1, Rd2 and Rd3 introduce compensating delay to the right signal pulse. Consequently, the left signal pulse will now coincide with one of the delayed right pulses and the corresponding AND circuit will then produce output. The operation is similar when the sound source is located to the left of the centreline. In that case the right signal is compared to the delayed left signal. The operation of the circuit is depicted in Figure 5.39.

In Figure 5.39 the left input pulse (L input) coincides with the delayed right pulse (R delay 2) and consequently the $Sd(2)$ signal is generated. The output of this direction detector is a pulse train. This can be transformed into a continuous signal by pulse stretching circuits.

The sound direction angle can also be estimated by the relative sound intensities at each ear or microphone. From Figure 5.36 it is easy to see that the right side intensity has its maximum value and the left side intensity has its minimum value when $\delta = 90°$. The right and left side intensities are equal when $\delta = 0°$. Figure 5.40 depicts the relative intensities of the right and left sides for all δ values.

It can be seen that the sound intensity difference gives unambiguous direction information only for the δ values of $-90°$ and $90°$ (the sound source is directly to

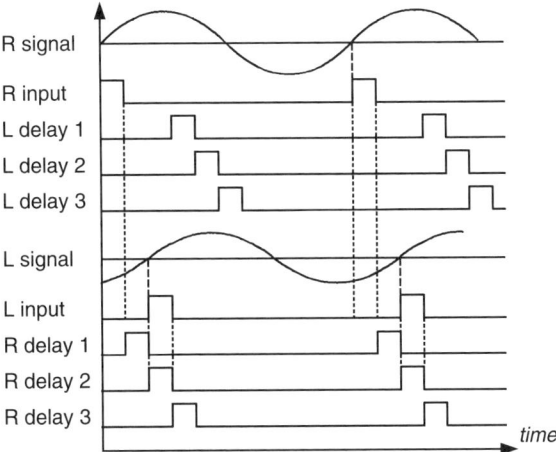

Figure 5.39 The operation of the arrival time comparison circuit

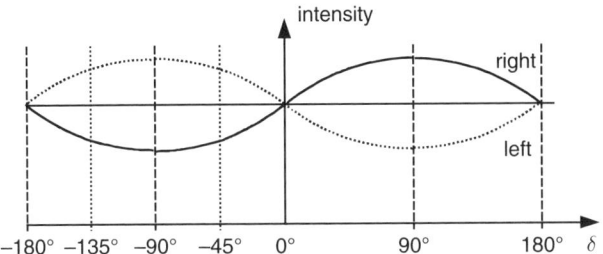

Figure 5.40 The sound intensities at the left and right ears as a function of the sound direction angle

the right and to the left). At other values of δ two directions are equally possible; for instance at $-45°$ and $-135°$ the sound intensity difference is the same and the sound source is either in front of or behind the head.

The actual sound intensity difference depends on the frequency of the sound. The maximum difference, up to 20 dB, occurs at the high end of the auditory frequency range, where the wavelength of the sound signal is small compared to the head dimensions. At mid frequencies an intensity difference of around 6 dB can be expected. At very low frequencies the head does not provide much attenuation and consequently the sound direction estimation by intensity difference is not very effective.

The actual sound intensity may vary; thus the absolute sound intensity difference is not really useful here. Therefore the comparison of the sound intensities must be relative, giving the difference as a fraction of the more intense sound. This kind of relative comparison can be performed by the circuit of Figure 5.41.

In the circuit of Figure 5.41 the direction is represented by the single signal vector $\{Sd(-3), Sd(-2), Sd(-1), Sd(0), Sd(1), Sd(2), Sd(3)\}$. In this circuit the right

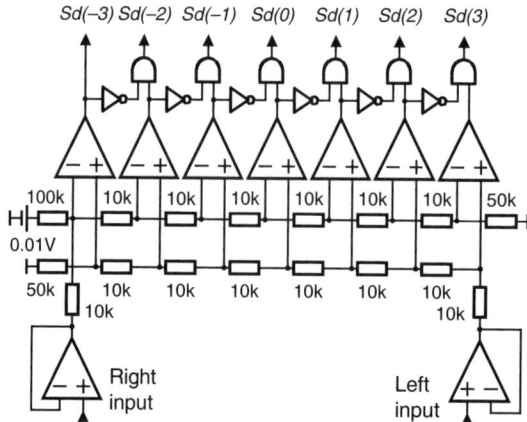

Figure 5.41 A relative intensity direction detector

and left input signals must reflect the average intensity of each auditory signal. This kind of signal can be achieved by rectification and smoothing by lowpass filtering.

In this circuit the left input signal intensity is compared to the maximally attenuated right signal. If the left signal is stronger than that the $Sd(3)$ output signal will be turned on. Then the slightly attenuated left input signal intensity is compared to a less attenuated right signal. If this left signal is still stronger than the right comparison value, the $Sd(2)$ output signal will be turned on and the $Sd(3)$ signal will be turned off. If this left signal is weaker than the right comparison value then the $Sd(2)$ signal would not be turned on and the $Sd(3)$ signal would remain on indicating the resolved direction, which would be to the right. Thus a very strong left signal would indicate that the sound source is to the left and the $Sd(-3)$ signal would be turned on and in a similar way a very strong right signal would indicate that the sound source is to the right and the $Sd(3)$ signal would be turned on. The small bias voltage of 0.01 V is used to secure that no output arises when there is no sound input.

These direction detectors give output signals that indicate the presence of a sound with a certain frequency in a certain direction. The intensity of the sound is not directly indicated and must be modulated on the signal separately.

Both direction estimation methods, the arrival time difference method and the relative intensity difference method, suffer from front–back ambiguity, which cannot be resolved without some additional information. Turning of the head can provide the additional information that resolves the front–back ambiguity. A typical front–back ambiguity situation is depicted in Figure 5.42, where it is assumed that the possible sound directions are represented by a number of neurons and their output signals. In Figure 5.42 the sound source is at the direction indicated by $Sd(2) = 1$. Due to the front–back ambiguity the signal $Sd(6)$ will also be activated. As a consequence a phantom direction percept may arise if the system were to direct the right ear towards the sound source; opposing head turn commands would be issued.

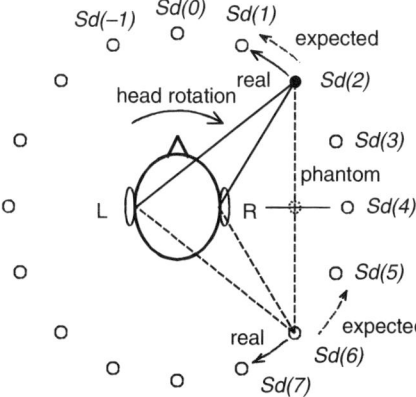

Figure 5.42 Head turning resolves the front–back ambiguity

The head may be turned in order to bring the sound source directly in front so that the intensity difference and arrival time difference would go to zero. In this case the head may be turned clockwise. Now both differences go towards zero as expected and $Sd(1)$ and finally $Sd(0)$ will be activated. If, however, the sound source had been behind the head then the turning would have increased the intensity and arrival time differences and in this way the true direction of the sound source would have been revealed.

5.8.7 Auditory motion detection

A robot should also be able to detect the motion of the external sound generating objects using available auditory cues. These cues include sound direction change, intensity change and the Doppler effect (frequency change). Thus change detection should be executed for each of these properties.

5.9 DIRECTION SENSING

The instantaneous position of a robot (or a human) is a vantage point from where the objects of the environment are seen at different directions. When the robot turns, the environment stays static, but the relative directions with respect to the robot change. However, the robot should be able to keep track of what is where, even for those objects that it can no longer see. Basically two possibilities are available. The robot may update the direction information for each object every time it turns. Alternatively, the robot may create an 'absolute' virtual reference direction frame that does not change when the robot turns. The objects in the environment would be mapped into this reference frame. Thereafter the robot would record only the direction of the robot against that reference frame while the directions towards the objects in the reference frame would not change. In both cases the robot must know how much it has turned.

112   MACHINE PERCEPTION

The human brain senses the turning of the head by the inner ear vestibular system, which is actually a kind of acceleration sensor, an accelerometer. The absolute amount of turning may be determined from the acceleration information by temporal integration. The virtual reference direction that is derived in this way is not absolutely accurate, but it seems to work satisfactorily if the directions are every now and then checked against the environment. The turning of the body is obviously referenced to the head direction, which in turn is referenced to the virtual reference direction.

Here a robot is assumed to have a turning head with a camera or two as well as two binaural microphones. The cameras may be turned with respect to the head so that the gaze direction will not always be the same as the head direction. The head direction must be determined with respect to an 'absolute' virtual reference direction. The body direction with respect to the head can be measured by a potentiometer, which is fixed to the body and the head.

For robots there are various technical possibilities for the generation of the virtual reference direction, such as the magnetic compass, gyroscopic systems and 'piezo gyro' systems. Except for the magnetic compass these systems do not directly provide an absolute reference direction. Instead, the reference direction must be initially set and maintained by integration of the acceleration information. Also occasional resetting of the reference direction by landmarks would be required. On the other hand, a magnetic compass can be used only where a suitable magnetic field exists, such as the earth's magnetic field.

The reference direction, which may be derived from any of these sensors located inside the head, must be represented in a suitable way. Here a 'virtual potentiometer' way of representation is utilized. The reference direction system is seen as a virtual potentiometer that is 'fixed' in the reference direction. The head of the robot is 'fixed' on the wiper of the virtual potentiometer so that whenever the head turns, the wiper turns too and the potentiometer outputs a voltage that is proportional to the deviation angle β of the head direction from the reference direction (Figure 5.43).

In Figure 5.43 the virtual reference direction is represented by the wiper position that outputs zero voltage. The angle β represents the deviation of the robot head

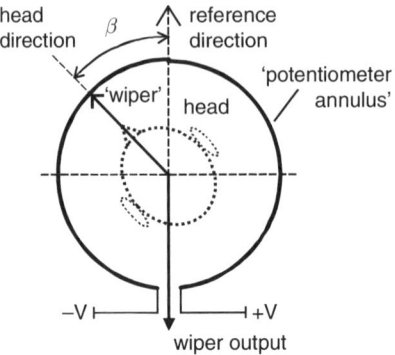

Figure 5.43   The head turn sensor as a 'virtual potentiometer'

direction from the reference direction. If the head direction points towards the left, the wiper output voltage will be increasingly negative; if the head direction is towards the right, the wiper output voltage will be increasingly positive. The angle value β and the corresponding virtual potentiometer output are determined by the accelerometer during the actual turning and are stored and made available until the head turns again. Thus the accelerometer system operates as if it were an actual potentiometer that is mechanically fixed to a solid reference frame.

5.10 CREATION OF MENTAL SCENES AND MAPS

The robot must know and remember the location of the objects of the environment when the object is not within the field of visual attention, in which case the direction must be evoked by the memory of the object. This must also work the other way around. The robot must be able to evoke 'images' (or some essential feature signals) of objects when their direction is given, for instance things behind the robot. This requirement can be satisfied by mental maps of the environment.

The creation of a mental map of the circular surroundings of a point-like observer calls for determination of directions – what is to be found in which direction. Initially the directions towards the objects in the environment are determined either visually (gaze direction) or auditorily (sound direction). However, the gaze direction and the sound direction are referenced to the head direction as described before. These directions are only valid as long as the head does not turn. Maps of surroundings call for object directions that survive the turnings of the robot and its head. Therefore 'absolute' direction sensing is required and the gaze and sound directions must be transformed into absolute directions with respect to the 'absolute' reference direction. Here the chosen model for the reference direction representation is the 'virtual potentiometer' as discussed before.

The absolute gaze direction can be determined from the absolute head direction, which in turn is given by the reference direction system. The absolute gaze direction will thus be represented as a virtual potentiometer voltage, which is proportional to the deviation of the gaze direction from the reference direction, as presented in Figure 5.44.

In Figure 5.44 β is the angle between the reference direction and the head direction, α is the angle between the gaze direction and the head direction and γ is the angle between the reference direction and the gaze direction. The angle α can be measured by a potentiometer that is fixed to the robot head and the camera so that the potentiometer wiper turns whenever the camera turns with respect to the robot head. This potentiometer outputs zero voltage when the gaze direction and the robot head direction coincide.

According to Figure 5.44 the absolute gaze direction γ with respect to the reference direction can be determined as follows:

$$\gamma = \alpha + \beta \qquad (5.21)$$

114 MACHINE PERCEPTION

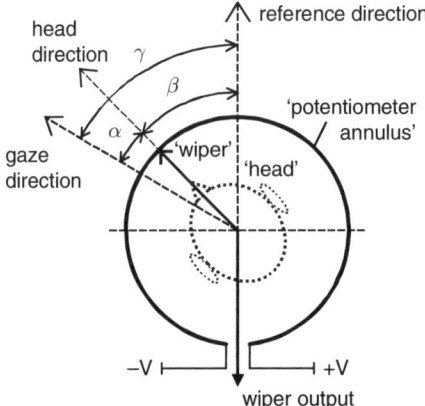

Figure 5.44 The determination of the absolute gaze direction

where the angles α and β are negative. Accordingly the actual gaze direction angle γ with respect to the reference direction is negative.

The absolute sound direction may be determined in a similar way. In this case the instantaneous sound direction is determined by the sound direction detectors with respect to the head direction (see Figure 5.36). The sound direction can be transformed into the corresponding absolute direction using the symbols of Figure 5.45:

$$\varphi = \delta + \beta \qquad (5.22)$$

where the angle δ is positive and the angle β is negative. In this case the actual sound direction angle φ with respect to the reference direction is positive.

Equations (5.21) and (5.22) can be realized by the circuitry of Figure 5.46. The output of the absolute direction virtual potentiometer is in the form of positive or negative voltage. The gaze and sound directions are in the form of single signal vectors and therefore are not directly compatible with the voltage representation.

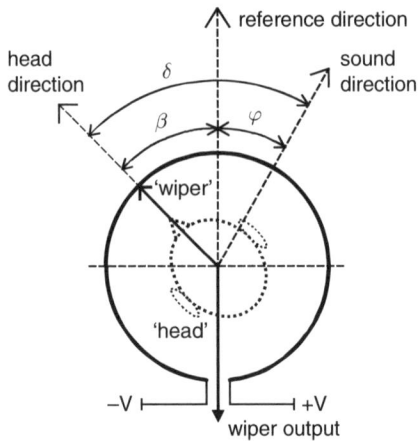

Figure 5.45 The determination of the absolute sound direction

CREATION OF MENTAL SCENES AND MAPS 115

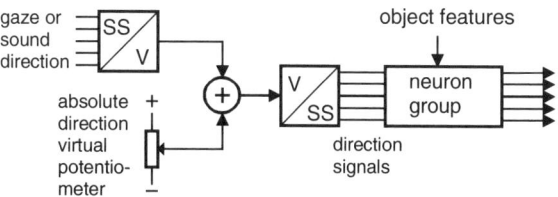

Figure 5.46 A circuit for the determination of the absolute gaze or sound direction

Therefore they must first be converted into a corresponding positive or negative voltage by a single signal/voltage (SS/V) converter. Thereafter the sums of Equations (5.21) or (5.22) can be determined by summing these voltages.

The sum voltage is not suitable for the neural processes and must therefore be converted into a single signal vector by the voltage/single signal (V/SS) converter. Now each single signal vector represents a discrete absolute gaze or sound direction and can be associated with the features of the corresponding object or sound. By cross-associating the direction vector with visual percepts of objects or percepts of sounds a surround map can be created. Here the location of an object can be evoked by the object features and for the other way round, a given direction can evoke the essential features of the corresponding object that has been associated with that direction and will thus be expected to be there.

6

Motor actions for robots

6.1 SENSORIMOTOR COORDINATION

Humans and animals can execute complex motor actions without any apparent computations. As soon as a nearby object is seen it can be grasped; the readiness to do so seems to arise as soon as the object is perceived. A motor act can also be imagined and subsequently executed, again without any apparent computational effort.

In robotic applications this kind of lucid readiness would be most useful. However, the existing robots usually do not work in that way; each motion has to be computed and more complex acts call for complicated numeric modelling.

Here a more natural way of motion control is outlined, one that realizes the immediate readiness to act as a response to a perceived situation and also allows the execution of 'imagined' acts. All this is to be executed without numeric computations.

6.2 BASIC MOTOR CONTROL

In many technical applications the position of a mechanical component has to be controlled accurately. For instance in CD players the laser head position must be exactly correct at every moment; the tracking error must be zero at all times. This calls for elaborate control circuitry. In robotic applications, however, this kind of accuracy is not always necessary. A robot arm must execute a motion that allows it to grasp an object. Here the trajectory of the arm is not important; the final position is. There is no need to define a trajectory for the arm and try to control the tracking error against this trajectory. Thus here the basic motor control problem reduces to the need to move or turn a mechanical part like a hand, head or other component from its present position to a new target position. This involves running one or more motors in the correct directions until the desired position has been achieved; that is the desired target position and the actual present position are the same. This is a rather trivial feedback control application and therefore does not pose any novel challenge. Here, however, the task is to interface this kind of feedback control circuitry with

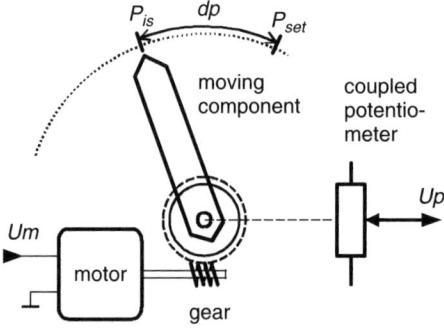

Figure 6.1 The basic motor control set-up

the associative neuron group system so that the requirements of the 'natural way of motion control' can be satisfied. In the following the principles that could be used towards this target are presented with examples of minimally complex circuitry. It is not claimed that these circuits would provide optimum control performance; the main purpose here is only to illuminate the basic requirements and principles involved.

As an introductory example a moving arm is considered. This arm is moved by a small permanent magnet electric motor via suitable gears. The position of the arm is measured by a potentiometer, which outputs a voltage that is proportional to the arm position (Figure 6.1).

In Figure 6.1 the present position of the arm is indicated as P_{is} and the desired target position is indicated as P_{set}. The potentiometer output gives the present position as a voltage value Up:

$$Up = c * P_{is} \tag{6.1}$$

where

$c =$ coefficient

The difference in the terms of the potentiometer voltage between the desired position and the present position is

$$du = c * (P_{set} - P_{is}) = c * dp \tag{6.2}$$

The polarity, value and duration of the motor drive voltage Um must now be administered so that the difference dp will go to zero. This can be achieved, for instance, by the following rules:

$$\text{IF} - Us < k * du < Us \text{ THEN } Um = k * du \text{ ELSE } Um = Us * du/|du| \tag{6.3}$$

where

Um = motor drive voltage, volts
P_{set} = desired position value
P_{is} = present position value
dp = position difference (mechanical position error)
du = position difference, volts
k = coefficient (gain)
Us = limiting value for motor voltage, volts

The control rule (6.3) can be graphically expressed as follows (Figure 6.2). In Figure 6.2 the motor drive voltage is given as the function of the error voltage that represents the position difference. When the position difference is zero, no correctional motion is necessary or desired and accordingly the motor drive voltage shall be zero. Whenever there is a positive or negative position difference, the motor must run to counteract this and bring the error voltage to zero. It is assumed that the circuitry is able to provide maximum plus and minus motor drive voltages ($+max$ and $-max$). The running speed of a permanent magnet DC motor is proportional to its drive voltage; therefore these voltage values would provide the maximum motion execution speed (and kinetic energy) for the mechanical system in both directions. However, it is desired that the execution speed were variable from a very slow execution to the maximum speed. Therefore the motor drive voltage is controlled by positive and negative limiting values $+Us$ and $-Us$, which are variable and controllable by external means.

The motor control characteristics of Figure 6.2 can be realized by the circuitry of Figure 6.3. (This circuit is presented only for illustrative purposes; a designer should consider the requirements of any actual application.) In this circuit the difference between the voltages that represent the desired position and the present position is computed by a differential amplifier. The difference or error voltage is then amplified by a gain stage. The amplified error voltage controls the final voltage follower power amplifier, which is able to supply enough current for the motor. The motor must be

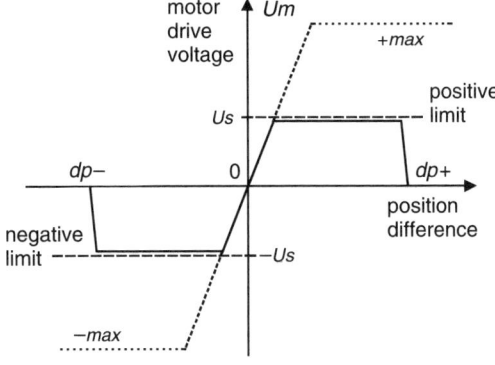

Figure 6.2 Motor control characteristics

120 MOTOR ACTIONS FOR ROBOTS

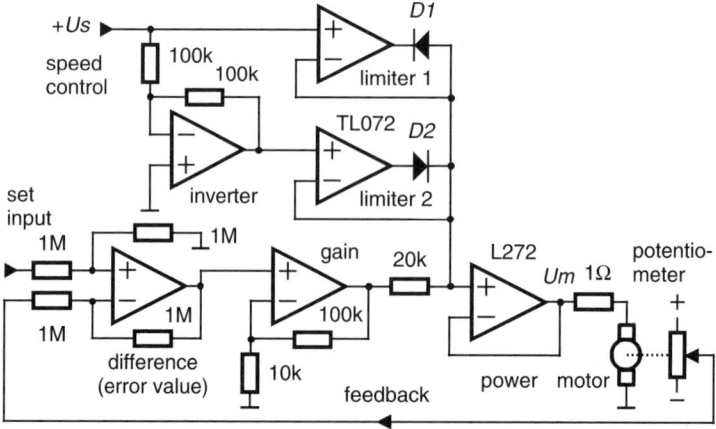

Figure 6.3 Motor control circuit for a small permanent magnet DC motor

of the permanent magnet type where the direction of rotation is determined by the drive voltage polarity. The motor polarity is to be observed; incorrect polarity will drive the motor towards the wrong direction and consequently the position error will grow instead of going to zero. In that case the wires going to the motor should be exchanged. This circuit requires a positive and a negative power supply.

Execution speed control is provided by the positive and negative limiter circuits that limit the amplified error voltage and the actual motor drive voltage, as the power amplifier is a voltage follower. The limiting values are controlled by the speed control input, which accepts controlling voltages between zero and a positive maximum value. The speed control may also be used to inhibit action by forcing the motor drive voltage to zero.

Feedback control loops usually incorporate some kind of lowpass filters in order to avoid instability. Here the motor mechanics have lowpass characteristics and accordingly no further lowpass elements are necessary. As a safety measure limit switches may be used. These switches would cut the motor drive-off if the allowable mechanical travel were exceeded. With proper component values this circuit works quite satisfactorily for small DC motors. However, for improved accuracy the benefits of PID (proportional-integral-derivative) controllers should be considered, but that is beyond the scope and purpose of this illustrative example.

6.3 HIERARCHICAL ASSOCIATIVE CONTROL

Next the interface between the associative system and the motor control system is considered. Assume that a desired target position for a moving mechanical component is acquired either via direct sensory perception or as an 'imagined' position. This position would be represented by a signal vector P. The motor control circuit of Figure 6.3 does not accept signal vector representations directly and, furthermore, the vector P would not directly represent position in motor control loop terms. The motor control loop is only able to drive the moving component to the position that

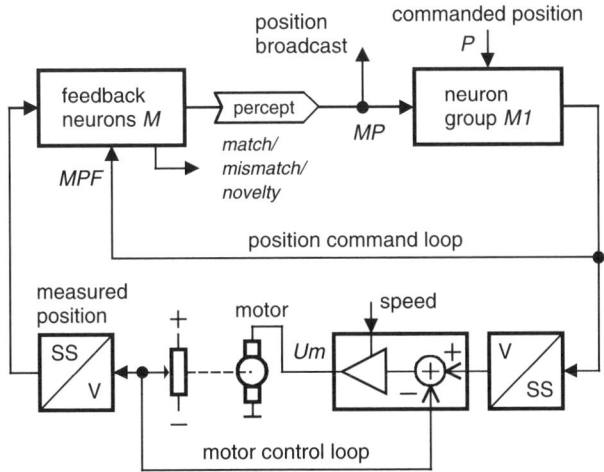

Figure 6.4 Movement control by the perception feedback loop

corresponds to the set input voltage. It is not practical that the overall associative system should know how to command and control the actual motor drive circuits in terms of their set value voltages. Instead, it is desired that the motor control circuits should be commanded by rather general commands. This can be achieved by the addition of hierarchical control loops, which accept associated commands. In Figure 6.4 one such realization is presented. The additional control loop, the position command loop, is in the form of a perception/response feedback loop.

In Figure 6.4 the motor and its control electronics are connected to a kinesthetic perception/response feedback loop that senses the position of the movable mechanical component by a potentiometer. The potentiometer voltage is transformed by the V/SS circuit into the corresponding single signal vector, which is accepted by the feedback neuron group M. The kinesthetic percept position single signal vector MP is broadcast to the rest of the system. The commanded position vector P evokes the corresponding kinesthetic single signal vector at the neuron group $M1$.

During an initial learning period the commanded position vectors P and the perceived kinesthetic position single signal vectors MP are associated with each other. When the learning is complete a commanded position vector P will able the corresponding kinesthetic single signal vector MPF to be evoked at the output of the neuron group $M1$.

The command vector P specifies only the requested end position of the moving part in terms of the requesting module, while the actual means of execution are not specified. That remains for the motor control loop, and the module that broadcasts the vector P does not know and does not have to know how the requested action will be executed.

In the motor control loop the vector MPF is transformed into the corresponding set-value voltage by the SS/V circuit. This causes the motor control feedback loop to drive the motor until the difference between the target voltage and the measured potentiometer voltage is zero. At that point the target position is achieved. Whether or not this act would actually be executed as well as the eventual execution speed would be determined by other means that control the motor speed.

The position command feedback loop must 'know' when the task has been executed correctly. This is effected by the match/mismatch/novelty detection at the feedback neuron group M, where the commanded position MPF is compared to the measured position. The resulting *match/mismatch/novelty* signals indicate the status of the operation. The *match* signal is generated when the reached position matches the command. The *mismatch* signal indicates that the commanded position has not been reached. The *novelty* signal arises when the measured position signal changes while no MPF signal is present. No motion has been commanded and thus the *novelty* signal indicates that the moving component has been moved by external forces.

6.4 GAZE DIRECTION CONTROL

Next these control principles are applied to control of the gaze direction in a simple robotic vision system. It is assumed that a digital camera is used as the visual sensor. It is also assumed that the camera can be turned along two axes, namely the pan (x direction) and tilt (y direction) axes by two separate motors. The visual sensor matrix of the camera is supposed to be divided into the high-resolution centre area and the low-resolution peripheral area, as described before. The peripheral area is supposed to be especially sensitive to change. The default assumption is that a visual change indicates something important, which would call for visual attention. Therefore a readiness should arise to allow the direction of the optical axis of the camera, the gaze direction, to be turned so that the visual change would be projected on to the high-resolution centre area (the fovea). This operation would lead to two consequences:

1. The visual change is projected to the main recognition area of the sensor.

2. The gaze direction indicates the direction of the visual change and is available as the instantaneous x, y values from the pan and tilt sensors (potentiometers).

Figure 6.5 illustrates the straight-ahead direction and the gaze and object direction in the x direction, where

α = angle between the straight-ahead direction and the gaze direction
ψ = angle between the gaze direction and the object direction
ω = angle between the object direction and the straight-ahead direction

These angles may have positive and negative values. It can be seen that the angle between the object direction and the straight-ahead direction is the sum of the angle between the straight-ahead direction and the gaze direction and the angle between the gaze direction and the object direction:

$$\omega = \alpha + \psi \qquad (6.4)$$

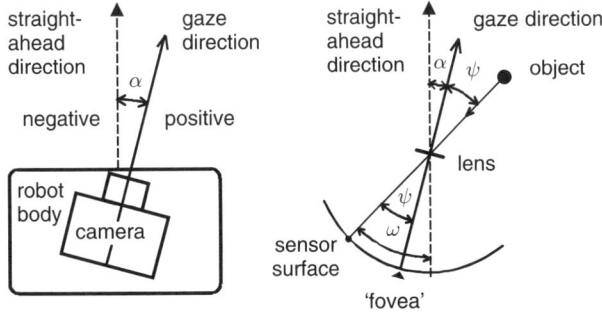

Figure 6.5 Straight-ahead direction, gaze direction and object direction

The angle ω represents the desired direction towards which the camera should be turned. This angle is independent of the actual gaze direction and depends only on the robot's instantaneous straight-ahead direction and the location of the object. When the gaze direction equals the object direction the ψ angle goes to zero and the α angle equals the ω angle.

For computation of the angle ω the angles α and ψ must be determined. The angle α may be determined by a potentiometer that is fixed on the body of the system. This potentiometer should output zero voltage when the camera points straight ahead, and a positive voltage when the camera points towards the right and a negative voltage when the camera points to the left. The angle ψ must be determined from the position of the projected image of the object on the photosensor matrix. This can be made with the aid of the temporal change detector (Figure 6.6).

The angle between the gaze direction and the object direction is reflected in the camera as the distance between the projected object image and the centre point on the photosensor matrix. The temporal change detector outputs one

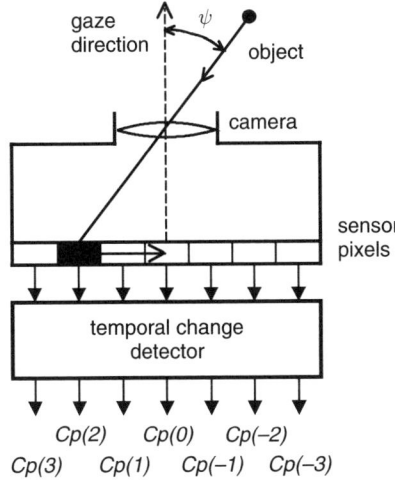

Figure 6.6 The camera with the temporal change detector

124 MOTOR ACTIONS FOR ROBOTS

active signal at a time if a visual change is detected. In Figure 6.6 the signals $Cp(3)$, $Cp(2)$, $Cp(1)$, $Cp(0)$, $Cp(-1)$, $Cp(-2)$, $Cp(-3)$ depict these signals. When a visual change is detected on the corresponding location, these signals turn from zero to a positive value. The location of the active Cp indicates the corresponding angle ψ.

A motor control circuit similar to the feedback control circuit of Figure 6.4 can be used for gaze direction control if the angle ω is taken as the commanded position input. For this purpose a circuit is added that determines the desired direction angle ω according to rule (6.4). The gaze direction motor control circuit is depicted in Figure 6.7.

The gaze direction control circuit of Figure 6.7 interfaces with the camera and temporal change detector of Figure 6.6 via a switched resistor network R_{-3}, R_{-2}, R_{-1}, R_1, R_2, R_3 and R. This network generates a voltage that is relative to the distance of each photosensor pixel location from the sensor matrix centre point and, accordingly, to the angle ψ. Each of the R_{-3}, R_{-2}, R_{-1}, R_1, R_2, R_3 resistors have their own switch, which is controlled by the corresponding Cp signal. Normally these switches are open and the voltage V_ψ is zero. An active $Cp(i)$ signal closes the corresponding switch and the voltage V_ψ will be determined by the voltage division by the resistors R_i and R. The signal $Cp(i)$ corresponds to a pixel location on the photosensor matrix and this in turn corresponds to a certain value of the angle ψ, which can be determined from the geometry of the system. The generated voltage V_ψ is made to be proportional to the angle ψ:

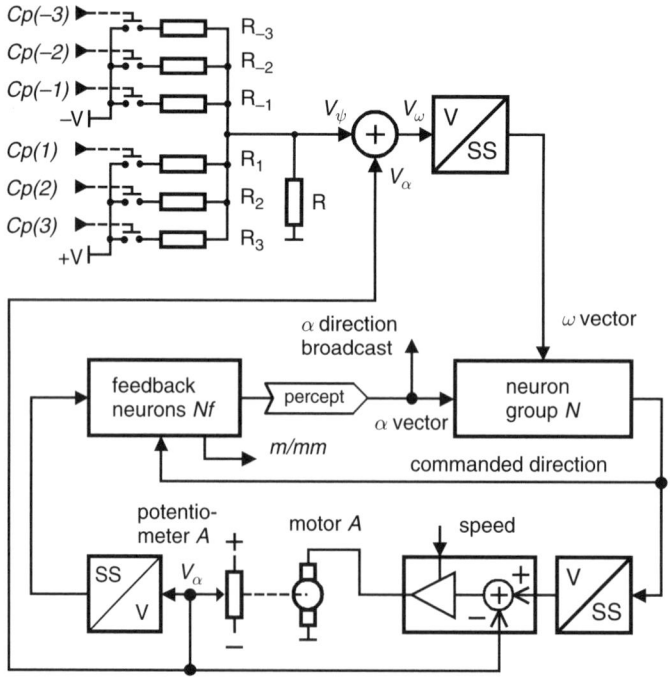

Figure 6.7 Gaze direction motor control circuit

$$V_\psi = V * R/(R+R_i) = k * \psi \qquad (6.5)$$

The required value for each resistor R_i corresponding to a given direction ψ can be determined as follows:

$$R_i = R * (V/(k * \psi) - 1) \qquad (6.6)$$

where

V = fixed voltage
k = coefficient

The signals $Cp(3)$, $Cp(2)$ and $Cp(1)$ correspond to positive ψ values and generate a positive voltage V_ψ while the signals $Cp(-1)$, $Cp(-2)$ and $Cp(-3)$ correspond to negative ψ values and generate a negative voltage V_ψ. The voltage V_ψ will be zero if the visual change is projected onto the centre point of the sensor matrix. In reality a large number of Cp signals would be necessary.

In Figure 6.7 the potentiometer A output voltage is

$$V_\alpha = k * \alpha \qquad (6.7)$$

From Equation (6.4),

$$V_\omega = V_\alpha + V_\psi \qquad (6.8)$$

During the desired condition the angle between the gaze direction and the object direction is zero:

$$V_\psi = 0 \qquad \text{(gaze is directed towards the object)}$$

During this condition, from Equation (6.8),

$$V_\alpha = V_\omega \qquad \text{(gaze direction equals the actual object direction)}$$

$$\alpha = \omega$$

Thus in the feedback control system of Figure 6.7 the V_ω value must be set as the SET value for the motor control loop. The V_ω value is computed according to Equation (6.8) by the summing element, as indicated in Figure 6.7. The V_ω value will not change when the camera turns. As continuous voltage values are not compatible with the neural system, the voltage/single signal transformation circuit is utilized. This circuit outputs the corresponding ω vector.

The ω vector evokes the corresponding α vector at the neuron group N. This vector is returned to the feedback neuron group Nf and, on the other hand, transformed into the corresponding voltage, which is the SET value for the motor control loop. The motor control loop will drive the motor to the positive or negative direction until the SET

126 MOTOR ACTIONS FOR ROBOTS

value and the measured V_α are equal. The *match/mismatch* signals from the feedback neuron group *Nf* indicate that the desired gaze direction is or is not achieved.

This system will turn the gaze towards the position of the visual change even if the change is discontinued and disappears. A complete gaze direction control system would have separate pan and tilt circuits that operate along these general principles. In a complete system a random scanning operation would also be included. This mode of operation could be used to find given objects and resolve their pan and tilt x, y location coordinates.

6.5 TRACKING GAZE WITH A ROBOTIC ARM

Next a more complicated motor action example is considered. Here a robotic arm with two joints is assumed. Each joint has its own potentiometer/motor combination, which is driven by the motor control circuitry as described before. Additionally it is assumed that the system has a visual sensor, which is able to pinpoint objects and give the corresponding x, y (pan and tilt) coordinates. The task for the robotic arm is to reach towards the visually pinpointed object and eventually grasp it. This system is depicted in Figure 6.8.

In Figure 6.8 the visual sensor detects the target object and determines its coordinates in terms of its own x, y coordinate pan and tilt system. The robotic arm also has its own coordinate system, namely the angle values ϑ and Θ for the joints. The working area and the maximum allowable values for the angles ϑ and Θ are limited so that unambiguous correspondence exists between each reachable $x \leftrightarrow (\vartheta, \Theta)$ and $y \leftrightarrow (\vartheta, \Theta)$.

The control system of Figure 6.4 can also be expanded to cover this application, so that detection of a target can immediately provide readiness to reach out and grasp it. This circuitry is depicted in Figure 6.9.

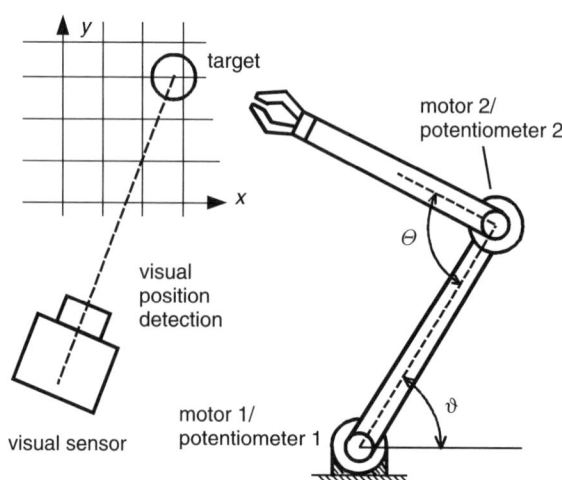

Figure 6.8 A robotic arm system with a visual sensor

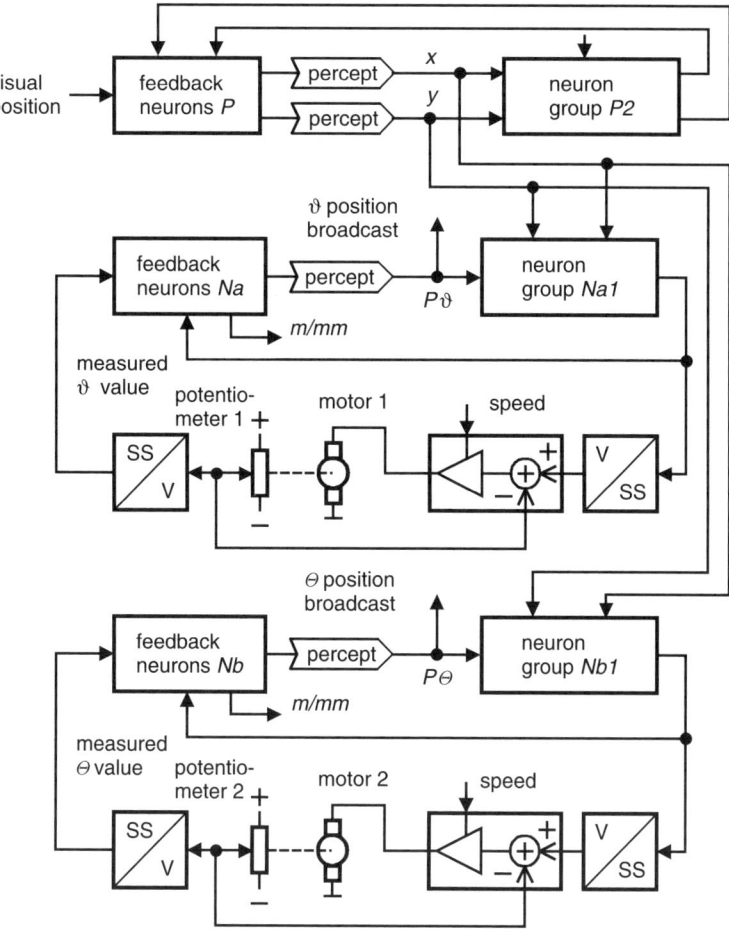

Figure 6.9 Circuitry for the robotic arm control

In Figure 6.9 the visual position signals x and y are determined by the visual perception/response loop and are broadcast to the two motor control circuits. The neuron groups $Na1$ and $Nb1$ translate the x and y signals into corresponding ϑ and Θ signals, which in turn are transformed into the corresponding target angle voltage values by the SS/V circuits. The motors 1 and 2 will run until the target angle values are achieved.

The weight values of the neuron groups $Na1$ and $Nb1$ may be determined by an initial training session, where each x, y location is associated with the corresponding ϑ, Θ angle values. On the other hand, this information can be deduced a priori from the geometry of the system and thus the weight values for the $Na1$ and $Nb1$ neuron groups may be 'programmed' in advance. However, if the system is designed to use tools or sticks that would extend the arm then the learning option would be useful.

Once properly trained in one way or an other the system will possess the immediate readiness to point out, reach out and touch the location that is indicated by the visual gaze direction. This will also work for imagined visual positions and in

128 MOTOR ACTIONS FOR ROBOTS

darkness. This readiness is immediate and fast as no computations are executed and no trajectories need to be considered.

This particular arm system has two joints, that is two degrees of freedom. When the joint angles are restricted, each target position will correspond only to one pair of angles. However, if the arm system had more joints, then there could be several sets of angle values for each target position. During learning some preferred sets of angles would be learned and the system would not normally use all the possible angle sets.

6.6 LEARNING MOTOR ACTION SEQUENCES

Many motor activities involve sequences of steps where during each step a mechanical component is moved from its present position into a new target position. An example of this kind of motor sequence would be the drawing of a figure with the robotic arm system of Figure 6.7. If the vision system scans a model drawing then the robotic arm will follow that and if fitted with a proper pen may copy the drawing. However, two shortcomings remain. Firstly, if the visual scanning is done too fast, then the arm may not be able to follow properly. Secondly, no matter how many times the operation is repeated the system will not be able to learn the task and will not be able to execute the task on command without renewed guidance. Therefore the basic circuitry must be augmented to allow the learning of motor sequences. The means to control the execution of individual steps must be provided; the next step must not be initiated if the previous step is not completed.

Figure 6.10 depicts the augmented circuitry. Here the motor system is again a part of a mechanical position perception/response loop. The position of the mechanical moving component is sensed by the coupled potentiometer as a continuous voltage value, which is then transformed into the corresponding single signal representation by the V/SS circuit. A possible target representation evokes the corresponding mechanical position representation at the output of the neuron group $M1$ in the same way as in the system of Figure 6.4. The neuron group $M2$ and the related shift registers R1, R2 and R3 form a sequence circuit similar to that of Figure 4.18.

The system shown in Figure 6.10 can be made to go through the steps of a sequence by showing the target positions in the correct order via the target representation input. The neuron group $M2$ utilizes correlative Hebbian learning, therefore some repetitions of the sequence are necessary. The sequence may be associated with an initiating command. Thus later on this command can be used to start the sequence. The sequence will then be completed without any further guidance from the system.

The execution of each consecutive step takes time and the next step cannot be initiated and the next mechanical target values should not be evoked until the present step is completed. In this circuit the timing is tied to the completion of each step. When the mechanical target position of a step is achieved, the match condition between the evoked target value (feedback) and the actual sensed value occurs at the feedback neuron group. The generated match signal is used as the timing signal

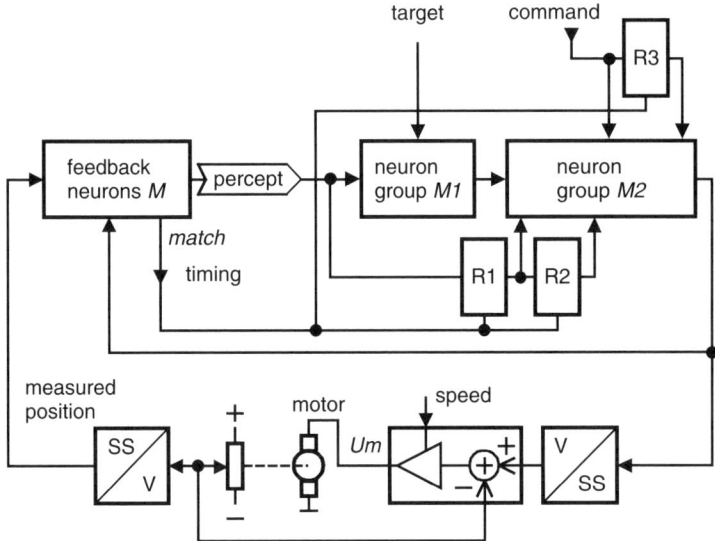

Figure 6.10 A simple system that learns motor action sequences

that advances the shift registers R1, R2 and R3. This causes the neuron group *M2* to output the next target position and the execution of the next step begins. The timing is thus determined by the actual mechanical execution speed. This in turn depends on the motor speed control input and any possible physical load.

In this example only one motor is used. It should be obvious that the same principles can be applied to systems with several motors and more complex mechanical routines.

6.7 DELAYED LEARNING

The learning of motor routines is controlled by feedback such as from the balance sensors and other sources. An action that disrupts balance is immediately seen as unfavourable and can be corrected, and the correct version of the motor sequences may be memorized during the execution. However, there are motor routines that can be judged only after the complete execution; an example of these kinds of routine is the game of darts. You aim and throw the dart and it is only after the dart has hit the board that you may know if your routine was correct. If the dart hits the bull's-eye the executed routine should be memorized, but this could only be done if the executed routine, the sequence of motor commands, is still available.

6.8 MOVING TOWARDS THE GAZE DIRECTION

A robot should be able to approach desired objects. In order to do so the robot should be able to determine the direction towards the desired object and then move towards that direction. A robot may locate an object visually; this visual location of the object

130 MOTOR ACTIONS FOR ROBOTS

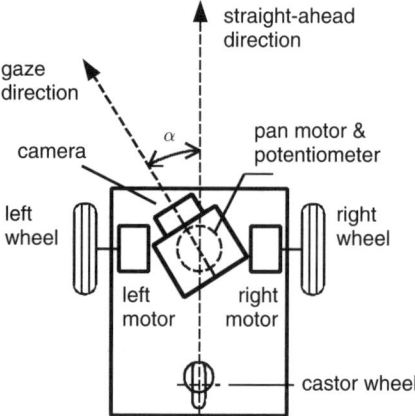

Figure 6.11 The arrangement for a simple robot (top view)

would involve the direction of the gaze towards the object, especially the x direction (the pan direction), as described before. Thereafter the gaze direction would equal the direction towards the object. The process of gaze direction controlled motion is illustrated here with the help of a simple hypothetical robot (Figure 6.11).

Figure 6.11 depicts a simple robot with three wheels and a moving camera. The camera can be turned towards given objects and this direction may be kept independent of the actual straight-ahead direction of the robot body. The robot is driven and turned by the left and right wheels and their motors. The robot turns to the left when the right wheel rotates faster than the left wheel and to the right when the left wheel rotates faster than the right wheel.

The task is to locate the desired object by panning the camera until the object is detected; thereafter the gaze direction will point towards the object. The robot should be turned now so that the gaze direction and the straight-ahead direction coincide. The angle α indicates the deviation between the gaze direction and the straight-ahead direction of the robot. This angle can be measured in terms of voltage by the pan-potentiometer. The potentiometer should be aligned so that whenever the angle α is zero the potentiometer output is zero. The right and left motors should be driven in a way that makes the robot turn towards the given direction so that the angle α eventually goes to zero. It is assumed that during this operation the camera and thus the gaze is directed towards the given direction constantly, for instance by means that are described in Section 6.4, 'Gaze direction control'. A simple drive circuit for the right and left motors is given in Figure 6.12.

In the circuit of Figure 6.12 the right and left motors have their own power amplifier drivers. A common positive or negative voltage, the *drive voltage*, causes the motors to run forward or backward, making the robot advance straight ahead or retreat. The pan-potentiometer is installed so that it outputs zero voltage when the gaze direction coincides with the robot's straight-ahead direction. Whenever the gaze direction is to the right the pan-potentiometer outputs a positive voltage. This voltage causes the left motor to run forwards (or faster forwards) and the right motor

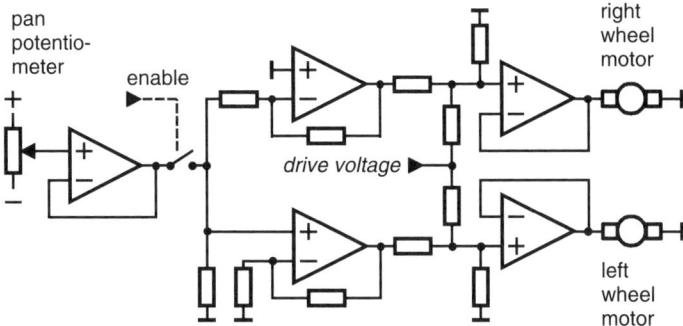

Figure 6.12 Drive circuit for wheel motors

backwards (or slower forwards, depending on the drive voltage) as the potentiometer voltage polarity is inverted in the right motor amplifier chain. This causes the robot to turn to the right until the potentiometer voltage goes to zero. Likewise, whenever the gaze direction is to the left the pan-potentiometer outputs a negative voltage. This voltage causes the left motor to run backwards and the right motor forwards and consequently the robot will turn to the left. The drive voltage will determine the actual forward motion speed of the robot. If the drive voltage is zero, the robot will turn on its place.

It is not useful for the robot to always turn towards the gaze direction; therefore an *enable function* may be provided. This function connects the pan-potentiometer output to the motor drive circuit only when the actual motor action is desired. The *enable function* would be controlled by the cognitive processes of the robot.

6.9 TASK EXECUTION

Next the combination of some of the previously described functionalities is outlined. Assume there is a small robot, like the one shown in Figure 6.11, and augment it with an auditory modality and a mechanical claw. Suppose further that the auditory modality allows the recognition of simple words and likewise the visual modality allows the recognition of certain objects. Thus, in principle, the robot could be verbally commanded to pick up certain objects like a soda can with a command like 'pick up (the) can'.

Commands like 'pick up can' do not describe how to locate the object, how to approach it or how to pick it up. These details remain to be solved by the robot. Therefore the robot must be able to look around and visually locate the can, it must move close enough and finally control its claw so that the pick-up action will be executed properly. The robot must be able to do all this on its own, without any further guidance from its human master. It should go without saying that in this case the robot must be able to do all this without microprocessors and preprogrammed algorithms. The hypothetical robot is depicted in Figure 6.13.

132 MOTOR ACTIONS FOR ROBOTS

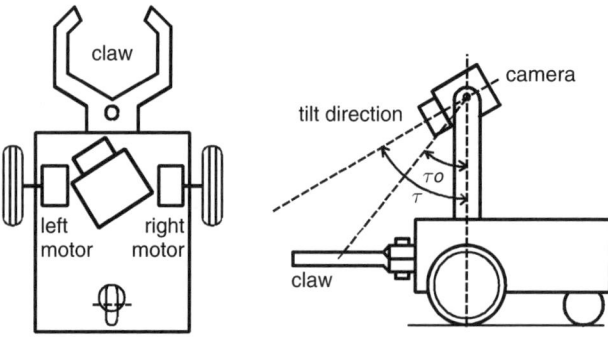

Figure 6.13 The robot with a camera and a claw

The robot shown in Figure 6.13 is driven by two wheels and wheel motors and is able to follow the gaze direction by means that were described earlier. In this way the robot may approach an object that is located in the gaze direction. However, the robot must also sense the straight-ahead distance to the object in order to be able to stop when the object is within the claw's reach. The straight-ahead distance to an object is estimated via the camera tilt (vertical gaze direction) angle τ. The angle τo corresponds to the distance where an object is directly within the claw (see also Figure 5.22). The camera tilt and pan directions are sensed by potentiometers.

The circuitry for the can-picking robot is outlined in Figure 6.14. The circuitry consists of the auditory perception/response loop, the visual perception/response loop, the camera pan and tilt motor loops, the claw motor loop, the system reaction unit and the wheel motor drive circuit, which is similar to that of Figure 6.12.

The command 'pick up can' is perceived by the auditory modality. The meaning for the word 'can' has been previously grounded to visual percepts of cans. Therefore the 'Accept-and-Hold' (AH) circuit of the neuron group $V2$ will accept the word 'can' and consequently evoke visual features of cans. If the robot does not see a can at that moment the mismatch condition will occur at the feedback neurons group V and a visual mismatch signal is generated. This signal is forwarded to the system reaction unit, which now generates a scan pattern to the pan and tilt motors as a hard-wired reaction to the visual mismatch condition. This causes the robot to scan the environment with the gaze. If a soda can is visually perceived the visual mismatch condition turns into the match condition and the scan operation ceases. Thereafter the wheel motor circuit is enabled to turn the robot towards the gaze direction.

The words 'pick up' relate to the claw motor circuit and basically cause the claw to close and lift, provided that the straight-ahead distance to the object to be picked up corresponds to the pick-up distance; that is the object is within the claw. Accordingly the words 'pick up' are accepted and held by the Accept-and-Hold circuit of the claw motor neuron group $C2$. The words 'pick up' are also accepted and held by the Accept-and-Hold circuit of the tilt motor neuron group $T2$ and

TASK EXECUTION   133

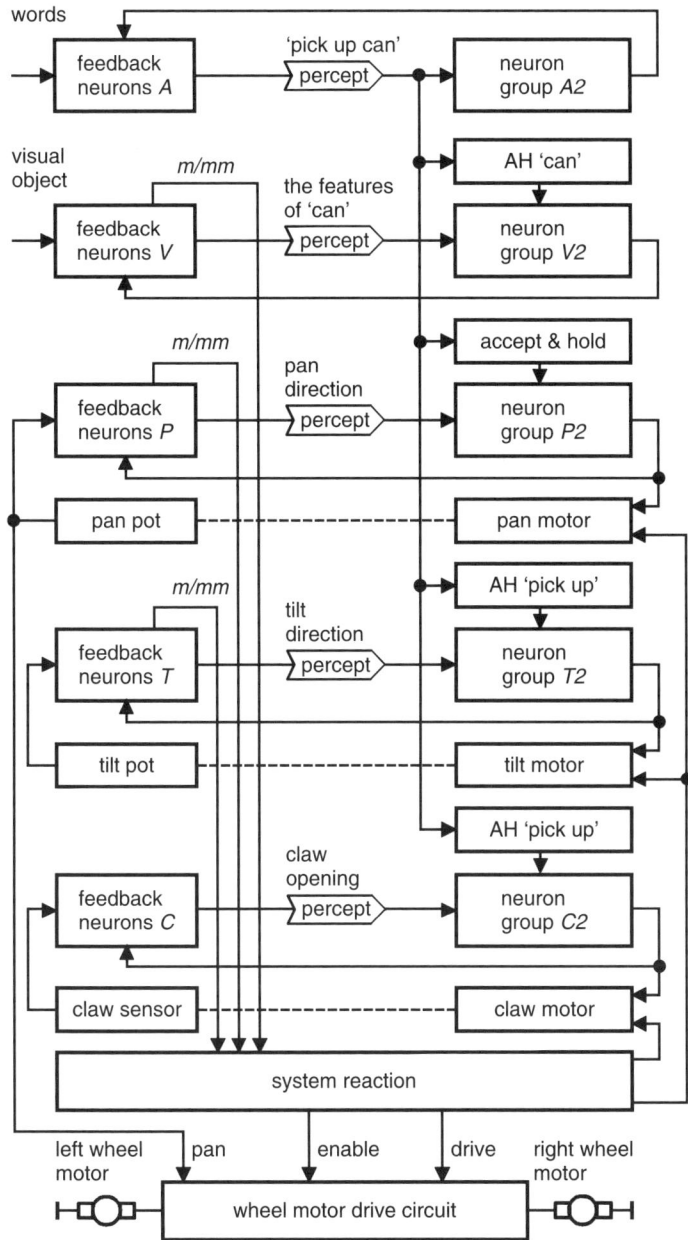

Figure 6.14 The circuitry for the can-picking robot

evoke there the corresponding pick-up angle τo. Thus, whenever the sensed tilt angle equals the angle value τo a tilt angle match signal is generated at the tilt motor feedback neuron group T. The simultaneous occurrence of the visual match signal for the desired object and the tilt angle match signal for the angle τo cause the hard-wired system reaction that closes the claw; the object will be picked up. Undesired objects that accidentally end up within the claw will not be picked up.

134 MOTOR ACTIONS FOR ROBOTS

The claw should be equipped with haptic pressure sensors, which would facilitate the limitation of the claw force to safe levels.

The behaviour of this robot is purely reactive. It executes its actions as reactions to the verbal commands and environmental situations. It does not imagine or plan its actions or predict the outcomes of these. It does not reason. It will not remember what it has already done. It does not have value systems or actual motives for action. The best that an autonomous purely reactive robot can do is to mindlessly wander around and do its act whenever a suitable situation arises. Obviously some enhancements are called for.

6.10 THE QUEST FOR COGNITIVE ROBOTS

Assume the existence of a rather simple home robot that is supposed to be able to execute simple everyday chores such as cleaning, answering the door, serving snacks and drinks, fetching items and the like. What would be the technical requirements for this kind of a robot and how could the previously presented principles be utilized here? What additional principles would be necessary? As an example a situation where the robot fetches some items and brings them to its master is considered (Figure 6.15).

Figure 6.15 depicts a modest living room where the master of the robot is watching television, a small kitchen and an adjoining room where the robot is located at the moment. On the master's command 'Robot, bring me a soda pop!' the home robot should go to the kitchen, take a glass out of the cupboard, go to the fridge and take a soda can and finally bring these to the table next to the chair where the master is sitting. Initially the robot is located in a position where it cannot see either the master or the kitchen.

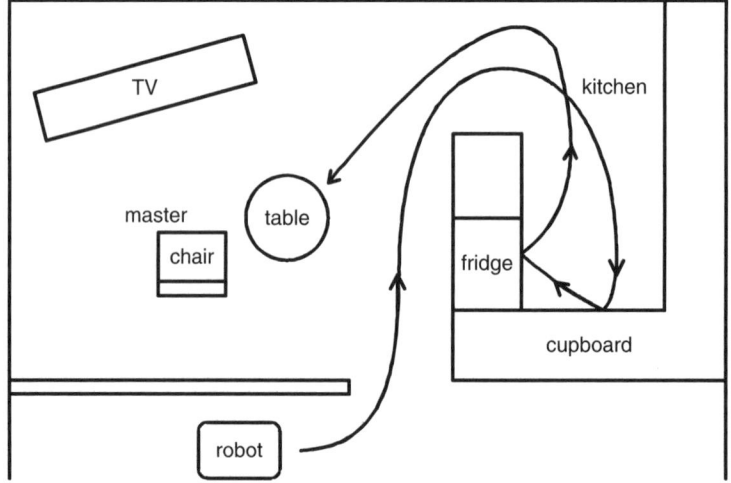

Figure 6.15 A task for a home robot

The robot's initial position excludes a direct reactive way of operation. Direct gaze contact cannot be made to any of the requested objects or positions and thus reactive actions cannot be evoked by the percepts of these. Therefore an indirect way of operation is called for. In lieu of the actual percepts of the requested objects the robot has to utilize indirect virtual ones; the robot must utilize mental representations of objects and environment based on earlier experience. In other words, the robot must 'imagine' the requested action in advance and execute action that is evoked by that 'imagery'. The path towards the kitchen cupboard would not be initiated by the gaze direction but by a mental map of the environment.

The necessary imagery should include a robot-centred 'mental map' of the environment, features of the cupboard, fridge, glass and the soda can. The command 'Robot, bring me a soda pop!' should evoke a mental sequence of actions where the robot 'sees' itself executing the command in its own mental imagery. This in turn would evoke the actual action that would be controlled by the percepts from the actual environment and the match/mismatch conditions between the imagined action and the actual situation. Thus, when the cupboard has been reached a match condition between the imagined location and the actual cupboard location would be generated; the robot would stop. This would allow the percept of the cupboard and the imagined percept of glasses inside to evoke the reactive action of opening the cupboard and the grasping of a glass. Likewise, the percept of the fridge and the imagined percept of a soda can inside would evoke the reactive actions of opening the fridge and the grasping of the soda can. Thus one purpose of the mental imagery would be to bring the robot to positions and situations where the desired reactive actions would be possible.

This example should show that even though reactive actions are fundamental and important, they are not sufficient alone for general task execution and must be augmented by other processes. These additional processes include imagination, mental reasoning and the ability to plan actions and to predict the outcome. These processes in turn call for the ability to remember and learn from experience and to create mental maps and models of objects, the environment and the robot self. These are cognitive processes; thus it can be seen that successful home or companion robots should be provided with cognitive powers.

7
Machine cognition

7.1 PERCEPTION, COGNITION, UNDERSTANDING AND MODELS

A robot that operates with reactive actions only does not have to know what it is doing; it just executes its actions as soon as the environment triggers them. Eventually this is not enough. Robots are needed that know what they are doing; cognitive robots are needed.

The term 'cognition' derives from the Latin word *cognoscere*, which means 'to know'. In cognitive psychology and neuroscience the term 'cognition' is usually used to refer to the processes of the mind, such as perception (which is discussed in Chapter 5, 'Machine Perception'), attention, reasoning, planning, imagination and memory. This kind of division is artificial, for in the brain these operations are not independent of each other and utilize the same processes and locations. Nevertheless, this division gives a general idea of the topic. Cognition may also be seen as a system's process of making sense: What is this all about and what should be done about it? Without renewing this implicit question continuously a cognitive being would not survive. This view of cognition links it with understanding and the mental means of achieving understanding, including reasoning and imagination.

Obviously answering the question 'What is this all about and what should be done about it?' begins with the determination of the meaning of the percepts of the situation. In the cognitive machine the direct meanings of percept signal vectors are hardwired to the point-of-origin sensory feature detectors and consequently the percept signal vectors indicate combinations of those features. The fact that sensor and feature detection errors may take place every now and then does not change this situation in principle. Thus, the percept vectors are just collections of feature signals selected by sensory attention and have no intrinsic meaning beyond their causal connection to the depicted features. For example, the meaning of an auditory feature signal may be 'the presence of a sound with a certain frequency'.

However, these kinds of intrinsic meaning of the detected visual and auditory feature patterns alone do not suffice for cognition and understanding. For instance, consider the ringing of a doorbell. Would it be possible to recover the meaning of the ringing of the doorbell by analysing the ringing sound carefully? No. The meaning must be associated with the sound via learning; that is the system must learn that the ringing indicates somebody being at the door. Thus the sensed feature patterns must be made to signify something via association. Seen objects should evoke possibilities for action and heard sounds should evoke ideas about the source and cause of the sound. Humans do not see and experience mere patterns of light intensities on the retina; they see *objects with additional meanings* out there. Likewise, humans do not hear and experience mere sounds; they hear *sounds of something*. Thus, the meaning of a sound goes beyond the intrinsic meaning of the stimulus as the sound pattern; the meaning is the cause of the sound, a telephone, a car, the opening of a door. This should go to show that simple perception in the form of pattern recognition and classification does not suffice for cognition; associated meanings are needed for understanding.

Modern cognitive psychology has proposed mental models as a device for cognition (Johnson-Laird, 1993) and for linguistic understanding (situation models) (Zwaan and Radvansky, 1998). There seems to be some experimental proof that humans indeed use mental models. The concept of mental models can also be utilized for perceptual cognition.

In this context a mental model is not like the numeric models used in electronic circuit simulation. Instead, a mental model is an associative network of representations that, for instance, allow the completion of an imperfect percept, the prediction of continuation, the representation of relationships and relative spatial locations, and the inferring of causes.

Understanding is complicated by the fact that simple perception processes produce ambiguous percepts. Individual percepts cannot always be taken at their face value; instead they must be interpreted so that they fit to each other. This fitting can be done efficiently in an indirect way by using mental models. Percepts are fitted to a mental model, which is evoked by the context, situation or expectation. If the fitting is successful, the situation is understood and all the background information that is available for the model can be used. These mental models may be simple, for instance for a human face, or they may be complicated and have a temporal episodic structure. These temporal models also relate to situational awareness. It is not proposed that these models were in the form of detailed imagery or a 'movie'; instead they may consist of minimal signal vectors that represent only some salient features of the model, such as position and some prominent property.

The combination of models that are evoked by the sensed environment and situation can be thought of as the system's running model of the world. This model is constantly compared to the sensory information about the world and match/mismatch conditions are generated accordingly. If the model and the external world match then the system is on track and has 'understood' its situation.

Advanced mental models also involve the perception of cause–consequence chains and the reasons and motives behind actions. A situation can have an explanation if the cause is revealed: 'The battery went dead because there was a short-circuit'. The explanation of things in terms of the already known mental model is understanding. The understanding of completely new matters calls for the creation of new coherent models. Faulty models lead to misunderstanding.

In information processing a model is an abstract and mathematically strict representation of entities with their properties, relationships and actions, and is suited for symbolic processing. Here, however, a model is seen as associatively accessible information about things, locations, actions, etc., with associated emotional significance. These kinds of model are not formally strict or complete and are suited for associative processing.

7.2 ATTENTION

A cognitive system cannot perceive everything at once and process all possible associations at the same time. Operations must be performed on selected items only, not on every object without discrimination. Therefore a cognitive system must have mechanisms that focus the perception and internal association processes on the most pertinent stimuli and context. This selection is called attention.

Sensory attention determines the focus of sensory perception. In the visual modality the focus of sensory attention is determined by the high-resolution centre area of the visual sensor (the fovea), while a visual change at the peripheral visual sensor area may direct the gaze towards that direction. In the auditory modality attention can be controlled by the selection of the sound direction and auditory features. It is also possible that a percept from one sensory modality determines the subsequent focus of attention for other sensory modalities. For instance, a sudden sound may direct visual attention towards the sound direction.

Inner attention determines the focus of the thought processes and selects inner activities for further processing. In imagination the inner (virtual) gaze direction is a useful tool for inner attention.

The outlined cognitive system does not have a centralized attention control unit. Instead, the processes of attention are distributed and operate via threshold circuits at various locations. These thresholds select the strongest signals only for further processing. This mechanism can be readily used for attention control; the signals to be attended are to be intensified so that they will be selected by the threshold circuits. Thus the strongest stimuli, the loudest sound, and the most intense light would then capture attention. Also the novelty of the stimuli, temporal or spatial change, can be translated into signal strength. Pertinence can be made to amplify the relevant signals and thus make them stronger.

In a modular system the threshold operation should be carried out at each separate module in a way that would allow the processing of some of the low-level signals without global activation. In certain cases this peripheral process could lead to elevated pertinence, which in turn could lead to the amplification of these signals and to global attention.

Emotional significance and match/mismatch/novelty conditions are important in attention control. These mechanisms are described in their corresponding chapters.

7.3 MAKING MEMORIES

7.3.1 Types of memories

Cognitive science divides memories into sensory memories, working memories, short-term memories, long-term memories, semantic memories and skill (or procedural) memories. Sensory memories hold sensory sensations for a while. In the visual modality the sensory memory is called the iconic memory; in the auditory modality it is called the echoic memory. The capacity of the iconic memory seems to be very small. On the other hand, the echoic memory allows the reproduction of a recent sound with high fidelity. Working memories hold a limited number of items that are relevant to the instantaneous cognitive task. Short-term memories (STMs) store the immediate history. Long-term memories (LTMs) are more permanent and store older history. (Some older textbooks use the term 'short-term memories' to mean the working memory function.) Semantic memories store general information such as 'a rose is a flower'. Skill memories store learned motor action routines. This division of memory types has been mainly based on memory function experiments and not so much on the actual material organization of the memory function in the brain, as this has been poorly understood.

It seems obvious that the aforementioned memory functions should also be implemented in a cognitive machine. On the other hand, the actual material division may not have to be the same as that in the brain.

The outlined cognitive system utilizes several kinds of memories. Sensory memories and working memories are realized with the basic structure of the perception/response loop and Accept-and-Hold circuits. Semantic memories are realized with the cross-connected neuron groups ($V1$, $A1$ groups, etc.). Skill memories are realized with the sequence neuron groups in the kinesthetic–motor perception/response loops, as described in Chapter 6. The realization and operation of short-term and long-term memories are described in the following.

7.3.2 Short-term memories

Short-term and long-term memories facilitate the retrieval of important information that is no longer available via sensory perception or is no longer sustained by the perception/response feedback loops. The memory function allows the cognitive agent to operate beyond the limits of immediate sensory perception and the limits of the present moment. The memory function facilitates thinking and the usage of mental models.

Short-term memories relate to recent situations, such as what happened recently, where was I, what did I do, where did I leave my things, etc. Short-term memories store practically everything that has been the focus of attention. The basic operation principle of the short-term memory function is described via a simple example 'where is my car' (Figure 7.1).

In Figure 7.1 four successive situations are presented. In the morning the car is at my home, at noon I am working and the car is at the company car park, in the

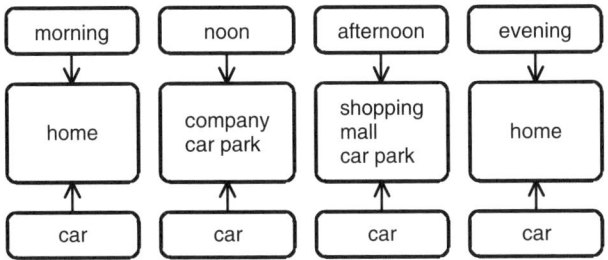

Figure 7.1 Recent memories: 'where is my car'

afternoon I visit the local shopping mall and my car is at the shopping mall car park and in the evening I am home again with my car.

What kind of memory system would be required and what kind of memories should be created that would allow a robot to operate successfully in this kind of a situation – to recall where the car was at each time? Obviously, the car, its location and the time point should be associated with each other. The time point may be represented by various means like percepts of sunrise, actual clock time, etc. The home and the car may be represented by some depictive features. It is assumed here that these representations for the time point, the car and the car park are based on visual percepts; thus the required association has to be done in the visual domain. For the act of association the signal vectors for these entities must be available simultaneously; therefore some short-term memory registers are required. One possible neural circuit for the short-term memory operation for the 'where is my car' example is presented in Figure 7.2.

In Figure 7.2 the Accept-and-Hold circuits AH1, AH2 and AH3 capture the signal vectors for the car, place and time. Thereafter these entities are cross-associated with each other by the neuron group *V2*. The synaptic weights arising from these cross-associations will constitute the memory traces for the situations <the car is at home in the morning>, <the car is at the company car park at noon>, <the car

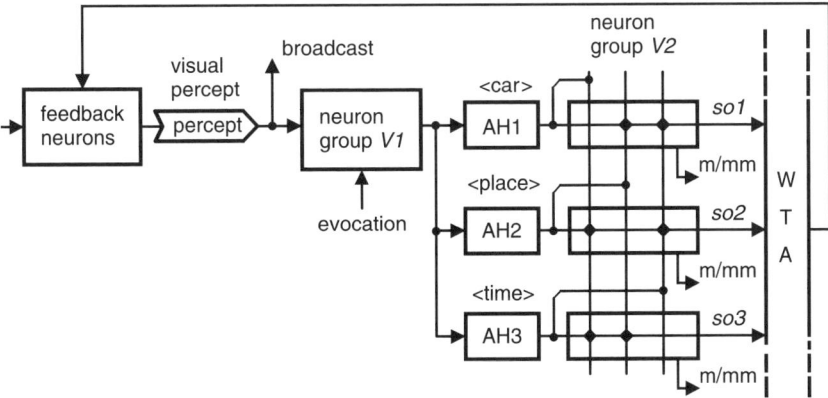

Figure 7.2 A neural circuit for the 'car park' example

is at the shopping mall car park in the afternoon> and <the car is at home in the evening>. After these associations the system is able to recall:

1. Where was the car at a given time?

2. When was the car at a given location?

3. What was at a given location at a given time (the car).

For instance, the recall of the imagery for the car and noon will evoke the signal vectors for <car> and <noon>, which will be captured by the Accept-and-Hold circuits AH1 and AH2. These will then evoke the response $so2$, the mental depiction for the <company car park> at the neuron group $V2$ via the two synaptic weights, as shown in Figure 7.2. The circuit will handle the other situations of Figure 7.1 in a similar way.

What would happen if a false claim about the whereabouts of the car at a given time were entered in the circuit? For instance, suppose that the claim that the car was at the shopping mall car park at noon is entered. Consequently the Accept-and-Hold circuit AH1 will accept the <car>, the AH2 circuit will accept the <shopping mall car park> and the AH3 circuit will accept the time <noon>. Now, however, the <car> and <noon> will evoke the location <company car park>, which will not match with the location hold by the AH2 circuit. Thus a mismatch signal is generated at the neuron group $V2$ and the system is able to deny the claim.

In this way the short-term memory system memorizes the relevant associative cross-connections as the situations change and accumulates a personal history of situations that can be later on associatively recalled with relevant cues whenever the past information is needed.

Figure 7.2 is drawn as if single signal (grandmother signal) representations were used. This is a possibility, but it is more likely that feature signal vectors would be used in practice. This does not affect the general principle.

7.3.3 Long-term memories

Most short-term memories tend to lose their relevance quite soon as time goes by. Therefore it would be economical to let irrelevant memories fade away to make room for more recent and probably more relevant memories. On the other hand, there are some memories that are worth retaining. These should be captured from the flow of short-term memories and stored in a way that would not interfere with the recent memories.

The importance of certain memories may manifest itself in two ways:

1. Important matters are circulated, 'rehearsed', in the perception/feedback loops and are recalled from the short-term memory and memorized again.

2. Important matters have high emotional significance.

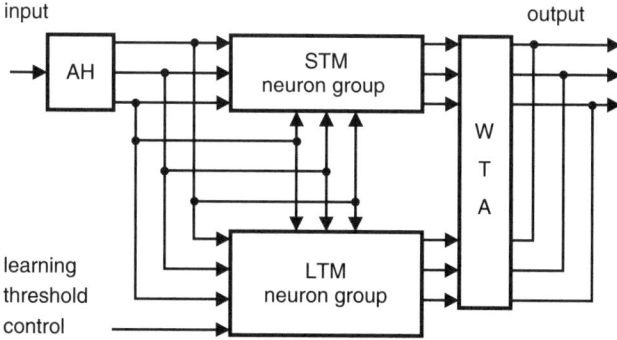

Figure 7.3 The addition of the long-term memory neuron group

Accordingly, the memorization of important matters can be implemented by neuron groups that utilize correlative Hebbian learning with an adjustable integration time constant. Normally many instants of association would be necessary and the matter would be memorized only after extended rehearsal. This process would also tend to filter out irrelevant associations, leaving only the repeating part for memorization. However, high emotional significance would lower the learning threshold so that the situation would be memorized without much filtering. Even the emotionally significant memories should first be captured by the short-term memory as the emotional significance of the event may reveal itself later.

In Figure 7.3 the STM neuron group is similar to the $V2$ neuron group in Figure 7.2. This neuron group learns fast but also forgets eventually. The LTM neuron group is connected in parallel with the STM neuron group and thus tries to do the same associations as the STM neuron group. However, the LTM neuron group utilizes slow correlative Hebbian learning and will therefore learn only frequently repeating connections between input entities. The learning threshold control input gets its input from the emotional significance evaluation circuit and allows fast learning.

7.4 THE PERCEPTION OF TIME

The human sense of passing time arises from the cognitive continuity provided by the short-term memory function. This is demonstrated by the unfortunate patients suffering from *anterograde amnesia*. This condition arises from a brain injury and is characterized by the inability to memorize and consciously remember anything that has taken place after the time of the injury. Examples of this condition are Korsakoff's syndrome (brain damage caused by alcohol), lesions due to brain infections and by-products of brain surgery. These patients seem to have a functional working memory with a very short time-span, but they are not able to form new memories and forget everything almost immediately. For them there is no temporal continuum, time does not go by and the present moment with a time-span of some tens of seconds is all they have. They cannot remember the recent past after the time of their injury nor plan for their actions (Nairne, 1997, p. 308; Rosenfield, 1995, Chapter III). In this way the scope of their consciousness is severely limited. The recallable long-term memories before the injury do not help to create a sense of passing time. The missing function is the creation of recent memories, that is the function of the short-term memory as it is defined here. Therefore, it can be seen that the short-term memory function should be essential for the creation of the sense of passing time and temporal continuum in a machine.

The temporal continuum includes the present moment, the past and the expected future. The present moment is different from the remembered past and the imagined future because only the present is grounded to real-time sensory percepts. The system is doing something right now because it gets sensory percepts about the action right now. The present moment is represented by the instantaneous sensory percepts and the present action is represented by the percepts from change detectors. For instance, a seen moving object generates visual change signals and a moving body part generates kinesthetic change signals. The system is 'experiencing' or 'living through' the action right now. On the other hand, memories and imaginations are represented by real-time percepts of the mental content, but the contents of memories and imaginations are not grounded to real-time sensory percepts. The system is 'experiencing' the act of remembering or imaging, but it is not 'experiencing' the contents of the memories or imaginations because the actual sensory signals are missing. The difference between remembered past and imagined future is that the system can remember some of the actual sensory percepts of the past while for the imagined future there are none.

The perception of the present moment can also be considered via the mental model concept. The complex of sensory percepts and the activated associative links constitute the 'running mental model' of the situation. This mental model is constantly matched against the sensory information and the match condition indicates that the model is valid. Memories and imaginations also involve mental models, but these do not match with instantaneous sensory percepts.

Another fundamental difference between the real present moment and the remembered or imagined moment can be hypothesized. The sense of self is grounded at each moment to sensory percepts, while the remembered or imagined self executing an action is not. Thus at each moment there should not be any doubt about the actual present moment self. The self in memories and imaginations is not the actual self that is here right now; it is only an 'image' of the self. This hypothesis leads to the question: What would happen to the sense of time and self if the grounding of the self to sensory percepts could somehow be cut off? Obviously the imagined self should replace the real self. This would seem to be so. During sleep self-related sensory percepts are practically cut off and, indeed, in dreams the imagined self is taken as the actual self. Consequently, the dreamer believes to live through the dream and the imagined moment is taken as the real present moment.

The perception of time also involves the ability to estimate the duration of short intervals. The proposed system contains circuitry that is able to recall and replay timed episodes such as music or motor sequences (see Chapter 4, Section 4.12, 'Timed sequence circuits'). This circuitry is able to time intervals and represent the timed duration by a signal vector. This signal vector can be stored in a short-term memory and may be used to reproduce the temporal duration.

Human perception of time includes the subjective feel of growing boredom when nothing happens. Should a machine have similar 'experiences'? Maybe.

7.5 IMAGINATION AND PLANNING

Machine imagination is defined here as: (a) the evocation of sensory representations without direct sensory stimuli depicting the same and (b) the manipulation of representations without direct sensory stimuli depicting the same (Haikonen, 2005b). Thus, the evocation of percept signals depicting an object in the visual perception/response loop would be counted as imagination if the system was not perceiving visually that object at that moment. This definition also allows verbally guided imagination like 'imagine a cat with stripes', as in this case no visual perception of a real cat would be present. It would also be possible to impose imagined features on real objects, such as 'imagine that this cat has stripes'. The second part of the definition relates to virtual actions with the percepts of imagined or real objects. For instance, actions that involve presently seen objects may be imagined. Transformations and novel combinations of imagined or real objects may be imagined. Imagination may involve locations that are different from the real location of the machine. Motor actions may also be imagined.

This definition connects imagination with perception. The sensors produce only a limited amount of information, which has to be augmented by imagination. It is not possible to see behind objects, but it can be imagined what is there. The outlined cognitive system is not necessarily a good recognizer; instead the percepts may evoke imaginations about the perceived objects, what they might be and what they might allow the machine to do. The mental models are augmented by imagination.

In the perception/response loop imagination is effected by the feedback that projects internally evoked signal vectors back to the percept points where they are treated as percepts. In this way imagined percepts will overlap with the actual sensory percepts, if there are any. How can the system separate percepts of imagination and percepts of the real world from each other? The real world consists of visual objects that overlap each other; likewise the auditory scene consists of overlapping sounds. A properly designed perception system must be able to deal with overlapping percepts. The imagined percepts are just another set of overlapping percepts to be dealt with.

Imagination can be triggered by sensory percepts and internal needs. A seen object may evoke the imagined act of grasping it, etc. Internal needs may include energy and environment management, such as energy replenishment, avoidance of extreme temperatures and wetness, etc. The environment and given tasks together may trigger imagery of actions to be executed.

Planning involves the imagination of actions that would lead to a desired outcome. In the course of planning, various sequences of action may be imagined and virtually executed. If these actions are similar to those that have been actually executed before then the outcome of the earlier action will be evoked and this outcome will be taken as the prediction for the outcome of the imagined action. If the outcome matched the desired outcome, then the imagined action could actually be executed. On the other hand, if the predicted outcome did not match the desired outcome then this action would not be executed and no real-world harm would have been done.

7.6 DEDUCTION AND REASONING

Some information is received in an indirect form and is only available for further processing via processes that are called deduction and reasoning. Formal logical reasoning operates with strict rules, which do not introduce any new information into the situation. In the following the neural realization of causal deduction and deduction by exclusion is presented. It will be seen that in associative neural networks these deduction methods are not as strict as their logical counterparts.

Causal deduction is based on the experience that certain events are caused by some other events; there are causes and consequences. For instance, it is known that if you are out while it rains you can get wet. It can be deduced from this experience that if somebody comes in with a wet overcoat, then it is raining outside. Here the cause is the rain and the consequence is the getting wet. This can be expressed formally as follows:

The premise: A causes B
The conclusions: 1a. If B then A
                 2a. If A then a possibility for B
                 3a. No A then no B
                 4a. No B then no conclusion

The conclusion 1a states that the consequence cannot appear without the cause. The conclusion 2a states that the presence of the cause allows the consequence, but the consequence does not necessarily appear. It may rain, but the person does not have to get wet. The conclusion 3a states that without the cause there will be no consequence. The conclusion 4a states that the absence of the consequence does not prove or disprove the presence of the cause. In human reasoning the conclusions 2a and 4a are easily misconstrued.

Next the realization of causal deduction by association is considered. Here an associative short-term memory network such as that of Figure 7.2 is assumed. This network associates the cause and the consequence as follows:

A cross-associated with B

After this association the following four situations that correspond to the four conclusions arise:

1b. B presented – evokes A
2b. A presented – evokes B

3b. A not presented – B not evoked
4b. B not presented – A not evoked

The situation 1b seems to work correctly. If you get wet while being out of doors then it rains. The situation 2b seems to be incorrect; however, in practice it is useful to become aware of the possible outcome even though this is not sure. If it rains, you may get wet. Also the situations 3b and 4b are usually rational in practical cases. Thus, the basic short-term memory circuit executes causal deduction in a way that is not in all cases strict in the logical sense, but may still be satisfactory in practice.

Another common reasoning method is the deduction by exclusion. In many cases there are possible situations that are mutually exclusive; if one situation is valid then the others are not. For instance, my keys are either in my pocket or on the table. My keys are in my pocket. Therefore my keys are not on the table. The simple case of deduction by exclusion can be represented formally as follows:

The premise: A is either B or C
The conclusions: 1c. If A is B then A is not C
                 2c. If A is C then A is not B
                 3c. If A is not B then A is C
                 4c. If A is not C then A is B

Example cases are as follows:

1d. My keys are in my pocket. Therefore my keys are not on the table.
2d. My keys are on the table. Therefore my keys are not in my pocket.
3d. My keys are not in my pocket. Therefore my keys are on the table.
4d. My keys are not on the table. Therefore my keys are in my pocket.

In an associative short-term memory network such as that of Figure 7.2 resolving the cases 1d and 2d is realized directly via match/mismatch detection after an observation. If the observation A is B (my keys are in my pocket) is memorized (keys and pocket are associated with each other) then the proposition A is C (my keys are on the table) will lead to mismatch between the evoked and proposed entities (pocket versus table). Therefore the second proposition will be rejected.

In the case 3c the associative system would look for the proposition B (my keys are in my pocket) and find that this proposition would contradict the observation. Thus mismatch would be generated and by the system reaction of mismatch attention would be focused on the other alternative, the table. The case 4c would be resolved in a similar way.

Another everyday example of reasoning involves the following premises and the conclusion:

The premises: A is (at) B
                B is (at) C
The conclusion: A is (at) C

An example case is as follows:

1e. My briefcase is in my car. My car is in the parking lot.
   Therefore my briefcase is in the parking lot.

This kind of reasoning can again be executed by the short-term memory network. The 'image' of the briefcase will evoke the 'image' of the car, the 'image' of the 'car' will evoke the 'image' of the parking lot, and that is where I should go for my briefcase.

These examples show a style of reasoning that does not rely on context-free formal rules and symbol manipulation. This is not high-level symbolic reasoning where knowledge is represented as statements within a language, but reasoning based on the use of memories and match/mismatch detection. This would seem to be rather similar to the way of reasoning that humans do naturally. Of course, it would be possible to teach formal reasoning rules to humans and to the cognitive machine as well so that actual rule-based reasoning could be executed in the style of formal logic and traditional artificial intelligence.

8
Machine emotions

8.1 INTRODUCTION

Would you like to be like the Sphinx that feels no pain or pleasure? Nothing feels good or bad, nothing brings you satisfaction, nothing motivates you to do anything, you would do what you do as you would be made to do it. This would be a life without emotions, the empty life of a zombie and sleepwalker – or a robot.

Should robots have emotions? Traditionally reason and emotions have been seen as the opposite; emotions do not and must not have any part in logical reasoning. However, the role of emotions in cognition is nowadays generally accepted. They are seen to be essential to attention, learning, motivation and judgement. The value of emotions has been pointed out by LeDoux (1996), Damasio (2003) and others. In machine cognition emotional significance is seen as guiding learning and decision making (Davis, 2000; Haikonen, 2002).

In psychology there are various theories about emotions – what they are and how they operate. According to everyday experience emotions seem to involve a triggering event that causes overlapping effects of physiological reactions, subjective feelings and cognitive evaluation. This is also proposed by the Schachter and Singer (1962) two-factor theory. Plutchik (1980) proposed that there are only eight basic emotions, and they are: acceptance, anger, anticipation, disgust, joy, fear, sadness and surprise. All the other emotions are supposed to be combinations of these and each emotion can exist in varying arousal or intensity levels. Unfortunately these and other theories of emotion offer only vague guidance to the designer of cognitive machines. Therefore the author has tried to condense the essence of these theories into a practical approach to machine emotions (Haikonen, 2003a). This approach is not necessarily psychologically accurate, but it is artificially implementable in a way that leads to useful system behaviour. This approach is presented later on.

There are some robots and toys that display outer expressions of emotions without actually having any real emotional states. It is obviously easier to design robots like these than to create machines that actually have inner processes and system reactions that correspond to emotions. However, some effort has been made to implement actual emotion-like processes in cognitive robots (Dodd and Gutierrez, 2005). In addition to their functional effects emotions also have certain subjective feelings; it feels like something to be in emotion. Should a robot also have these subjective feelings and, if so, how could these be implemented? What would it take to make a machine really *feel* pain? What would it take to make a machine really *feel* pleasure? Would it be possible that some complex analog feedback control loop systems already feel pain, but have no way of communicating this fact to us?

Here the following aspects of emotions are considered: emotional significance evaluation, emotions as attention control, emotional states as templates for responses and emotions as motivational factors.

8.2 EMOTIONAL SIGNIFICANCE

An autonomous robot must be able to make decisions without continuous help from a human supervisor. Some decisions may be based on simple rules, while others may require more general criteria, possibly in the form of a value system. All decision events cannot be directly covered by preprogrammed rules. A robot must be able to think and reason on its own, plan and imagine alternative courses of action and evaluate the goodness or badness of the probable results of these. Would the predicted outcome match the expectation? Would the planned action result in a destructive and painful (whatever that would be in robot terms) outcome? Obviously the robot should not actually execute actions that could lead to undesired outcomes. A robot might learn to assess its imagined plans via experience and training.

Humans learn via the pleasure and pain, rewards and punishment that are caused by the event itself or by a human teacher. These emotional sensations mark the tried action as suitable or not suitable. Emotional markers help also to recognize events that call for immediate attention and fast responses in order to avoid major damage. These kinds of emotional marker are memorized along the actual events and form a kind of 'emotional soundtrack'.

It is proposed that robots should have a similar emotional significance system and an 'emotional soundtrack'. For this purpose a true cognitive robot should have the concepts of good, bad, pain and pleasure. The brain derives these concepts from elementary sensations like taste, smell, pain and pleasure and generalize these to apply to more abstract matters. It is proposed that a cognitive machine should derive these concepts in a similar way from elementary sensory information originating from suitable sensors. These sensors could include smell and taste as well as pain and pleasure. Even though a robot may not need to accept or reject things by their smell and taste, artificial sensors could nevertheless be used as good and bad value input points. In robotic applications physical damage sensors should be used as pain sensors. These inputs could then also be used to punish and reward the system.

8.3 PAIN AND PLEASURE AS SYSTEM REACTIONS

What would it take to perceive and feel pain and pleasure? Could it be reproduced artificially in a robot? What kind of a sensor could sense pain? In humans the meanings of the neural signals from the eyes are grounded to the seen objects of the outside world. These signals represent the sensed external entities. However, the feel of pain is not grounded in this way to sensed entities because pain is not a property of a sensed entity. Pain sensors do not sense pain. The sensed entity is cell damage and the generated neural signal commands the system to pay attention to this and react urgently. The pain signals do not carry the *feel* of pain; they only evoke a number of system reactions that may continue beyond the duration of the acute cause of the pain. These system reactions are related to the feel of pain. System reactions are not representations, and thus the feel of pain is not either.

The nonrepresentational nature of pain is also obvious from the fact that humans cannot memorize the *feel* of pain and evoke it afterwards as any other memory. Humans can remember that they had a headache, but this memory does not, luckily, include the *feel* of the headache. Likewise, pleasure is not a representation either, but a system reaction. A cognitive robot should utilize a similar pain/pleasure principle.

Pain signals indicate that something is wrong and the situation should not be continued. Pain signals alone do not usually tell what exactly should be done in order to remedy the situation. Therefore an array of general responses are launched. Some of the pain-related responses are: capture of attention, withdrawal, rejection, discontinuation of action, association of a 'bad' value with the action, avoiding the associated action in the future, aggression, retaliation, rest. It can be seen that these are not representations; these are actions and, more accurately, system reactions to the pain signals.

In a similar way, pleasure signals indicate that the ongoing action is favourable and should be continued. Accordingly, the pleasure-related responses include: fixation of attention, approaching, accepting, continuation of action, intensification of a related action, association of a 'good' value with the action, seeking the associated action in the future.

Here it is useful to notice that the effects of the match and mismatch conditions are somewhat similar to those of pleasure and pain. Both the match condition and pleasure try to sustain the existing focus of attention; both the mismatch condition and pain call for the redistribution of attention. Thus the concepts 'match pleasure' and 'mismatch displeasure' could be used and the pleasure and displeasure would be defined here via their functional effects.

Functional pain and pleasure can be realized in a machine via system reactions that produce the consequential effects of pain and pleasure. These reactions must be triggered by something. Therefore humans need 'pain' and 'pleasure' sensors, which provide the hardwired grounding of meaning for pain and pleasure as well as for goodness and badness. Match/mismatch detection is also necessary. In this way a machine can be built that reacts to, say, mechanical damage as if it were in pain; it will withdraw from the damage-causing act and will learn to avoid similar situations in the future. The machine may also try to use force to eliminate the damage-causing agent. 'Pleasure' may be related to energy replenishment, etc. Again, the machine would act as if it were experiencing pleasure. At this moment this is sufficient. The question 'Does the machine really feel pain?' relates to the question of consciousness and will be discussed in that context.

8.4 OPERATION OF THE EMOTIONAL SOUNDTRACK

The 'emotional soundtrack' contains the emotional significance of percepts and perceptual episodes and allows the emotional judgement of these and similar percepts as soon as they are evoked, either by sensory stimuli or as memories. Emotional evaluation can only be based on experience, the past connections between percepts and simultaneously occurring sensations of pain and pleasure. Thus, in the simplest realization the 'emotional soundtrack' is created via the association of the pain signals and pleasure signals with the simultaneously active percepts (Figure 8.1).

In Figure 8.1 the S vector represents a sensory feature array and the corresponding percept vector is Sp. The emotional soundtrack is created by the association of percept vectors Sp with the pain and pleasure signals at the displeasure and pleasure neuron groups. This association takes place whenever signals from pain or pleasure sensors are present.

Assume that a percept vector Sp has no association with pleasure while a pleasure sensor emits a signal. This signal goes through the pleasure neuron group. The output of this neuron group is the pleasure signal pls. Initially this signal is not associated with the Sp vector, but after a short while the associations take place at the pleasure neuron group and the neuron group S. Thereafter the intensity of the output F of the neuron group S will be elevated due to the associative evocation by the pls signal, as the intensity of the neuron group output signal is the sum of the direct signal and the evoked signal. This elevation will, in turn, intensify the percept Sp signals via the feedback loop. By the functional definition, perceived pleasure should try to endorse the pleasure-producing activity by sustaining attention on the same. Attention, on the other hand, is controlled by signal intensity. Thus the pleasure signal should intensify the broadcast percepts that are related to the ongoing activity. It can be seen that this process is achieved here.

Next, assume that a percept vector Sp has no association with pain while a pain sensor emits a signal. This signal goes through the displeasure neuron group and appears as the pain signal p. It is important that the pain signal is able to evoke the functional effects of pain immediately, without delay. The pain must try to

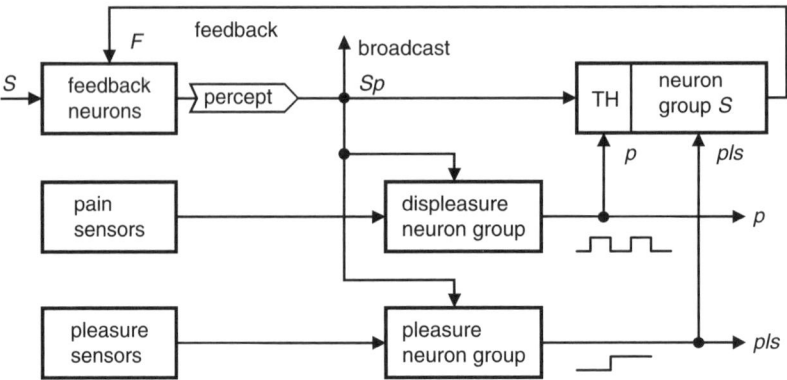

Figure 8.1 Emotional evaluation and the emotional soundtrack

stop whatever activities are going on. Therefore the pain signal should not rely on associative connections. Accordingly, in Figure 8.1 the pain signal is used to elevate the input threshold level of the neuron group S. This will lower the Sp percept signal intensity as described in Chapter 5 (see Figure 5.4). Eventually the pain signal and the Sp vector are associated with each other at the displeasure neuron group. Thereafter the p signal intensity is elevated, raising the neuron group S input threshold further. This in turn will lower the intensity of the percept vector Sp, which now, in turn, will lower the p signal intensity. This will then allow the Sp signal intensity to recover and the oscillatory cycle repeats itself. The advantage of this kind of operation would be that activities are not prevented completely. Competing activities may have a chance and eventually remedial activities, if available at all, could win.

It can be speculated that from the system's phenomenal point of view, if there is any, this kind of disruption of attention might not feel nice. However, it is, after all, supposed to be displeasure and pain.

8.5 EMOTIONAL DECISION MAKING

Artificial emotional decision making is here based on three ideas. Firstly, mental ideas have emotional values (the 'emotional sound track'), which are evoked if the idea is evoked. Secondly, a proposition is intensified or attenuated by the emotional values of the ideas that the proposition evokes. Thirdly, a proposition will initialize action if its intensity exceeds the execution threshold. Figure 8.2 depicts a simple example.

In Figure 8.2 the proposition 'should I go to the movies' evokes a number of ideas as a response, like 'I feel like going, there is a good movie', 'movies are fun', 'it is raining out there, I don't want to go out' and 'I am tired, I don't feel like going anywhere'. Each of these ideas carries emotional significance, which will affect the eventual decision about going to the movies. 'Good movie' and

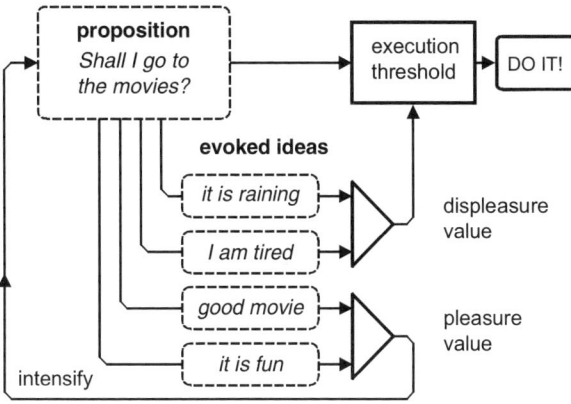

Figure 8.2 Emotional decision making

154 MACHINE EMOTIONS

'movies are fun' evoke pleasure, which according to the earlier definition will try to endorse the ongoing activity by elevating the intensities of the related signals. On the other hand, 'raining' and 'I am tired' evoke displeasure, which again by the earlier definition will try to suppress the proposed activity. If the signal intensity for the proposed action exceeds a certain execution threshold then the action will be executed; otherwise the proposition will fade away.

Emotional decision making is based on the agent's values and as a process is, in fact, quite rational. However, skewed values may lead to improper decisions.

8.6 THE SYSTEM REACTIONS THEORY OF EMOTIONS

8.6.1 Representational and nonrepresentational modes of operation

The system reactions theory of emotions (SRTE) for machines (Haikonen, 2003a) considers a cognitive machine as a dynamic system with representational and non-representational modes of operation. In addition to the associative processing of the representational signal vectors the system is assumed to have certain basic system reactions that relate to attention control and motor activity. These reactions are triggered and controlled directly by certain elementary sensor percepts and by the emotional evaluation of sensory and introspective percepts. In this kind of a system not only the contents of the internal representations matter but also the way they emerge and stay within the focus of attention. Thus two parallel processes may be triggered, one that is representational and leads to a cognitive report, and possibly also to actions, and another that leads to emotional evaluation, system reactions and system percepts of these reactions (Figure 8.3). These two processes are connected. A trigger may be an elementary sensation or a percept.

In Figure 8.3 the emotional process affects the cognitive process via basic circuit mechanisms such as threshold modulation. The cognitive process may include the self-reflective effects of inner speech – thoughts about one's emotional states like 'Am I now angry or what'. These in turn will be emotionally evaluated and may consequently alter the emotional state.

The elementary sensations <good>, <bad>, <pain>, <pleasure>, <match>, <mismatch> and <novelty> relate to system reactions that are hardwired into the

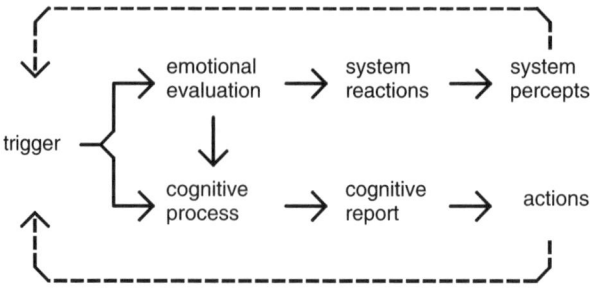

Figure 8.3 The system reactions theory of emotions (SRTE) model

Table 8.1 Elementary sensations, system reactions and typical motor functions

Elementary sensation	System reaction	Motor function
Good	Approach, accept	Forward
Bad	Withdraw, reject	Reverse
Pain, self-inflicted	Withdraw, discontinue	Fast reverse
Pain, external causes	Escape	Fast
	Aggression, attack	High force
Pain, overpowering	Submission, guard	Lock, freeze
Pleasure	Sustain, approach	Continue
Match	Sustain attention	
Mismatch	Refocus attention	
Novelty	Focus attention	Forward, slow

cognitive system. The system reactions for each elementary sensation are summarized in Table 8.1. This table considers the elementary sensations at a functional level. The subjective 'feel' of these or the lack of it is not considered at this moment.

The actual form of the system reactions depends on the machinery, its possible mechanical responses and degrees of freedom. The controllable motor functions that relate to these are the direction of action (forward, reverse) and motor speed from zero to a maximum value (execution speed with effects on force and kinetic energy). Attention is controlled by various threshold values and signal intensity. Sensory attention is partly controlled by the direction of the sensors (visual sensors, auditory sensors).

8.6.2 Emotions as combinations of system reactions

The system reactions theory of emotions proposes that combinations of system reactions lead to dynamic machine behaviour that corresponds to human emotions. Some emotions and their proposed corresponding system reaction combinations are given in Table 8.2.

Emotional system reactions manifest themselves as typical behaviour. Curiosity would appear as the attention fixation on novel stimuli and potentially as approaching the cause of the stimuli with explorative actions. Fear would appear as the avoidance and fleeing of the fear-causing stimuli. Desire-related emotions like love and affection would involve seeking the closeness to the object of the emotion and complying with its needs. This emotion would be useful for servant robots.

Emotional system reactions also have a temporal aspect. Astonishment would involve a large mismatch that is caused by the sudden failure of the system's running world model. Disappointment would involve the failure to gain an expected reward.

More complex behaviour would arise from conflicting emotions and motives. For instance, a given task might involve approaching a fear-evoking entity. In this case

Table 8.2 Emotions as combinations of system reactions

Emotion	System reactions of
Curiosity	Novelty + good
Astonishment	Mismatch (sudden large)
Fear	Bad + pain
Desire	Good + pleasure
Sadness	Mismatch + overpowering pain
Anger	Aggression
Disgust	Bad (intensive)
Caution	Novelty + good + bad

the motor commands to approach and to escape would conflict and might result in an oscillatory forward–reverse motion. The generated self-reports and the emotional evaluation of these would complicate the situation further.

8.6.3 The external expressions of emotions

In human interpersonal transactions it is useful to have some idea about the emotional state of others so that one's behaviour and attitude towards others may be modified accordingly. This goes for emotional robots as well. If a robot utilizes emotional criteria in its operation then it would be useful for the human master to have some indication about the emotional state of the robot at each moment. Humans convey information about emotional states via facial expressions, which are usually readily understood. Thus these kinds of facial expressions would also be useful for robot–human communications.

8.7 MACHINE MOTIVATION AND WILLED ACTIONS

Digital computers do what they do because they are programmed to do it; the programs force the execution of the specified actions. Also the IF-THEN-ELSE type of branching in a program code is not genuine decision making, but a programmer's way of specifying what the computer has to do in various situations. The computer does not make a decision here.

A true cognitive machine is not governed by a program. It has the capacity to learn and execute certain, hopefully useful, actions that it can execute on command, but it should also be able to do this on its own initiative, as it deems suitable.

Curiosity should be the first 'emotion' and motivation when the cognitive robot is switched on for the first time. By definition 'curiosity' is evoked by the perception of novel objects and these would be abundant for the robot initially. 'Curiosity' should lead to cautious examination of the robot itself and the environment, and in the course of this study the robot should be able to create its first inner models of the world and its own mechanical body.

Additional motivation is generated via pain and pleasure. Humans do something because it gives them pleasure or because it helps to avoid pain. This fundamental motivation mechanism can be applied to cognitive machines as well. Due to the basic system reactions a cognitive machine will strive towards 'pleasure'-producing actions and tries to discontinue and avoid 'pain'-producing actions. The emotional evaluation process associates these actions with pleasure and displeasure values, thus creating the 'emotional soundtrack' for these. Thereafter these values are evoked whenever the actions are imagined or suggested by the environment. The master of the machine may use the emotional significance as a motivational factor. The desired activities should be associated with 'pleasure' and the undesired actions should be associated with 'pain'. For this purpose the machine should have suitable 'pleasure' and 'pain' sensors that act as gateways to the pleasure and pain system reactions. Bump and collision sensors may be used as 'pain' sensors as they should, by their very nature, indicate nondesired incidences. In this way the basic system reactions and their effect on attention can be made to direct the machine towards desired actions.

'Pain' in the machine is related to physical damage. The machine will learn to expect 'pain' as the outcome of certain situations and will consequently try to avoid these situations. This behaviour corresponds to the self-preservation instinct.

Should the machine want something? To want something is to be in a situation where the desired situation mismatches the existing situation. This mismatch refocuses attention towards actions that try to realize the desired situation. The objects of desire are those that create expectations of pleasure. The state of wanting something is the precursor for the execution of the related act and as such a necessary state for a cognitive machine that is motivated by pleasure and displeasure. This is related to the concept of machine willed actions. A machine may have desired actions that it wants to execute in the previous sense, and consequently it will seek to do whatever may facilitate the execution of the action.

What should an idle cognitive robot do? The robot may have some given tasks to do whenever suitable situations arise, for instance cleaning and picking up trash, etc. These actions would be triggered by the environment. Other triggers could be percepts of task-related objects, an event, a given time. Sometimes, however, the environment may not readily give suitable stimuli. For those cases a basic 'emotion' should be provided, namely 'boredom'. In this state the machine would recall memories of pleasant acts and see if any of those could be executable now.

9

Natural language in robot brains

9.1 MACHINE UNDERSTANDING OF LANGUAGE

Practical interactive robots should be able to understand language. They should be able to discuss their situation, understand commands and explain what they have done and why, and what they are going to do. They should be able to learn by verbal descriptions. Companion robots should also be able to do small talk. Human thinking is characterized by silent inner speech, which is also a kind of rehearsal for overt spoken speech. A robot that uses language in a natural way may also have or need to have this inner speech.

Traditional natural language processing theories have treated languages as self-sufficient systems with very little consideration about the interaction between real-world entities and the elements of the language. However, in cognitive systems and robotics this interaction would be the very purpose of language. A robot will not understand verbal commands if it cannot associate these with real-world situations. Therefore, the essential point of machine linguistics is the treatment of the grounding of meaning for the elements of the language, the words and syntax, as this would provide the required bridge between language and the world entities. Thus the development of machine understanding of language should begin with consideration of the mechanisms that associate meaning with percepts.

The cognitive processes give meaning to the percept signal vectors that in themselves represent only combinations of features. Thus humans do not see only patterns of light; they see objects. Likewise they do not only hear patterns of sounds; instead, they hear sounds of something and infer the origin and cause of the sound.

However, language necessitates one step further. A heard word would be useless if it could be taken as a sound pattern only; it would be equally useless if it were to mean only its physical cause, the speaker. A seen letter would likewise be useless if it could be taken only as a visual pattern. Words and letters must depict something beyond their sensed appearance and immediate cause. Higher cognition and thinking can only arise if entities and, from the technical point of view, signal vectors that represent these entities can be made to stand for something that they are not.

How does the brain do it? What does it take to be able to associate additional meanings with, say, a sound pattern percept? It is rather clear that a sound and its source can be associated with each other, but what kind of extra mechanism is needed for the association of an unrelated entity with an arbitrary sound pattern? No such mechanism is necessary. On the contrary, additional mechanisms would be needed to ensure that only those sounds that are generated by an entity were associated with it. The associative process tends to associate everything with everything, whatever entities appear simultaneously. Thus sounds that are not causally related to an entity may be associated with it anyway. Consequently sound patterns, words, that actually have no natural connection to visually perceived objects and acts or percepts from other sensory modalities may nevertheless be associated with these and vice versa. An auditory subsystem, which is designed to be able to handle temporal sound patterns, will also be able to handle words, their sequences and their associations, as will be shown.

However, language is more than simple labelling of objects and actions; it is more than a collection of words with associated meanings. The world has order, both spatial and temporal, and the linguistic description of this order calls for syntactic devices, such as word order and inflection.

Communication is often seen as the main purpose of language. According to Shannon's communication theory, communication has been successful if the transmitted message can be recovered without error so that the received pattern is the exact copy of the transmitted pattern, while the actual meaning of the pattern does not matter. It is also important in human communication to hear correctly what others say, but it is even more important to understand what is being said. Thus it can be seen that Shannon's theory of communication deals only with a limited aspect of human communication (for example see Wiio, 1996). In human communication meaning cannot be omitted. A natural language is a method for the description of meaning and the use of language in human communication can be condensed as follows:

sensorily perceived situation → *linguistic description* → *imagined situation*

and

imagined situation → *linguistic description* → *imagined situation*

Thus a given perceived situation is translated into the corresponding linguistic description and this in turn should evoke imagery (in a very broad sense) of the corresponding situation, a situation model, in the receiver's mind. A similar idea has also been presented by Zwaan (2004), who proposes that the goal of language comprehension is the construction of a mental representation of the referential situation.

Obviously the creation of the imagined situation that corresponds to a given linguistic description cannot take place if the parties do not have common meanings for the words and a common syntax. However, even this may not be sufficient. Language provides incomplete descriptions and correct understanding calls for the inclusion of context and common background knowledge, a common model for the world.

In the following these fundamental principles are applied to machine use of natural language. Suitable representations for words are first considered, then a possible architecture for speech acquisition is presented and finally the 'multimodal model of language' is proposed as the 'engine' for machine use and understanding of natural language.

9.2 THE REPRESENTATION OF WORDS

In theory the heard words are sequences of phonemes. Therefore it would suffice to recognize these phonemes and represent each of them by a single signal. If the number of possible phonemes is, say, 26, then each phoneme could be represented by a signal vector with 26 individual signals, of which only one could be nonzero at any time. Most words have more than one phoneme; therefore words would be represented by mixed signal representations of $n*m$ signals, where n is the number of phonemes in the word and m is the number of possible phonemes (for instance 26).

A spoken word is a temporal sequence where the phonemes appear one at a time and are not available at the same time. Therefore the serial representation of a word must be transformed into a parallel form where all phoneme signals of a word are available at the same time. This can be done by the methods that are described in Chapter 4, 'Circuit Assemblies'.

In practice it may be difficult to dissect heard words into individual phonemes. Therefore syllables might be used. Thus each word would be represented by one or more syllables, which in turn would be represented by single signal vectors. The sequence of syllables must also be transformed into parallel form for further processing. For keyboard input the phoneme recognition problem does not exist, as the individual letters are readily available.

Still another possibility is to use single signal representation (grandmother signals) for each and every word. In this case the number of signals in the word signal vector would be the number of all possible words. This will actually work in limited applications for modestly inflected languages such as English. For highly inflected languages such as Finnish this method is not very suitable, as the number of possible words is practically infinite. In those cases the syllable coding method would be more suitable.

9.3 SPEECH ACQUISITION

Speech acquisition begins with the imitation of sounds and words. Children learn to pronounce words by imitating those uttered by other people. This is possible because their biological capacity is similar to that of other people; thus they can reproduce the word-sounds that they hear. In this way children acquire the vocabulary for speech. The acquisition of meaning for these words takes place simultaneously, but by different mechanisms.

Obviously it could be possible to devise robots that acquire their speech vocabulary in some other way; in fact this is how robots do it today. However, the natural way might also be more advantageous here.

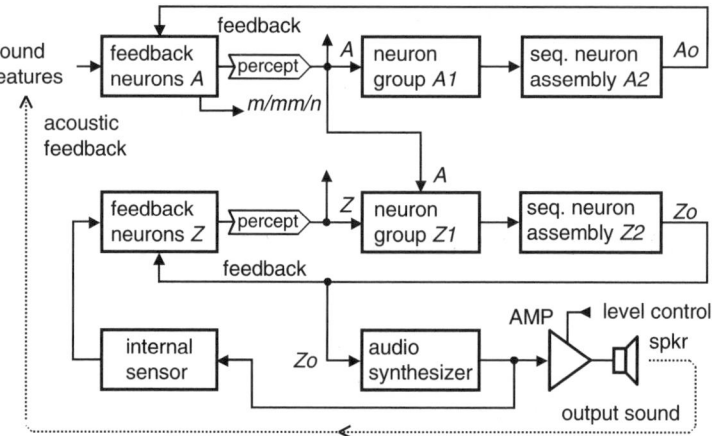

Figure 9.1 System that learns to imitate sounds and words

Sound imitation calls for the ability to perceive the sounds to be imitated and the capability to produce similar sounds. An audio synthesizer can be devised to produce sounds that are similar to the perceived ones. The synthesizer must be linked to the auditory perception process in a way that allows the evocation of a sound by the percept of a similar one.

The system of Figure 9.1 contains two perception/response loops, one for auditory perception and another for the kinesthetic perception loop for the control of the audio synthesizer output. The auditory perception/response loop receives the temporal sequence of sound features from a suitable preprocessing circuitry. The sequence neuron assembly $A2$ is used here as an echoic memory with instantaneous learning and is able to learn timed sound feature sequences. This circuit is functionally similar to that of Figure 4.31.

The audio synthesizer is controlled by the output signal vector Zo from the kinesthetic perception/response loop. The audio synthesizer output signal is forwarded to an audio amplifier (AMP) and a loudspeaker (spkr) via a controllable threshold circuit. It is assumed that the sound synthesizer is designed to be able to produce a large variety of simple sounds, sound primitives, and each of these primitives can be produced by the excitation of a respective signal vector Zo.

Assume that the sound producing signal vectors Zo were initially excited randomly so that each sound primitive would be produced in turn. These sounds would be coupled to the auditory sensors externally via the acoustic feedback and be perceived by the system as the corresponding auditory features A. These percepts would then be broadcast to the audio synthesizer loop and would be associated with the causing Z vector there at the neuron group $Z1$. This vector Z would then emerge as the output vector Zo. (The neuron groups $Z1$ and $Z2$ could be understood as 'mirror neurons' as they would reflect and imitate auditory stimuli. It should be evident that nothing special is involved in this kind of 'mirroring'.)

In this way a link between perceived sounds and the corresponding Zo vector signals would be formed. Later on any perceived sound would associatively evoke the corresponding *Zo* vector signals and these in turn would cause the production of a similar sound. This sound would be reproduced via the loudspeaker if enabled by the output level control. The system has now the readiness to imitate simple sounds.

Temporal sound patterns such as melodies and spoken words are sequences of instantaneous sounds. Complete words should be imitated only after they have been completely heard. The echoic memory is able to store perceived auditory sequences and can replay these. These sequences may then be imitated feature by feature in the correct sequential order one or more times. This kind of rehearsal may then allow the permanent learning of the sequence at the sequence neuron group $Z2$.

9.4 THE MULTIMODAL MODEL OF LANGUAGE

9.4.1 Overview

The multimodal model of language (Haikonen, 2003a) tries to integrate language, sensory perception, imagination and motor responses into a seamless system that allows interaction between these modalities in both ways. This approach would allow machine understanding of natural languages and consequently easy conversation with robots and cognitive systems.

The multimodal model of language is based on the assumption that a neuron group assembly, a plane, exists for each sensory modality. These planes store and associatively manipulate representations of their own kind; the visual plane can handle representations of visual objects, the auditory plane can handle representations of sound patterns, the touch plane can handle representations of tactile percepts, etc. The representations within a plane can be associated with each other and also with representations in other planes. It is also assumed that a linguistic plane emerges within the auditory plane, as proposed earlier. The representations, words, of this plane are associated with representations within the other planes and thus gain secondary meanings. According to this view language is not a separate independent faculty; instead it is deeply interconnected with the other structures and processes of the brain.

Figure 9.2 illustrates the idea of the multimodal model of language. Here five planes are depicted, namely the linguistic (auditory), visual, taste, motor and emotional planes. Other planes may also exist. Each plane receives sensory information from the corresponding sensors. The emotional plane would mainly use the pain and pleasure sensors. The representations within each modality plane may have horizontal connections to other representations within the same plane and also vertical connections to other modality planes. These connections should be seen changing over time. The temporal relationships of the representations can also be represented and these representations again can be connected to other representations, both horizontally and vertically. Representations that are produced in one plane will activate associated representations on other relevant planes. For instance, a linguistic sentence will activate representations of corresponding entities and their relationships on the other planes and, vice versa, representations

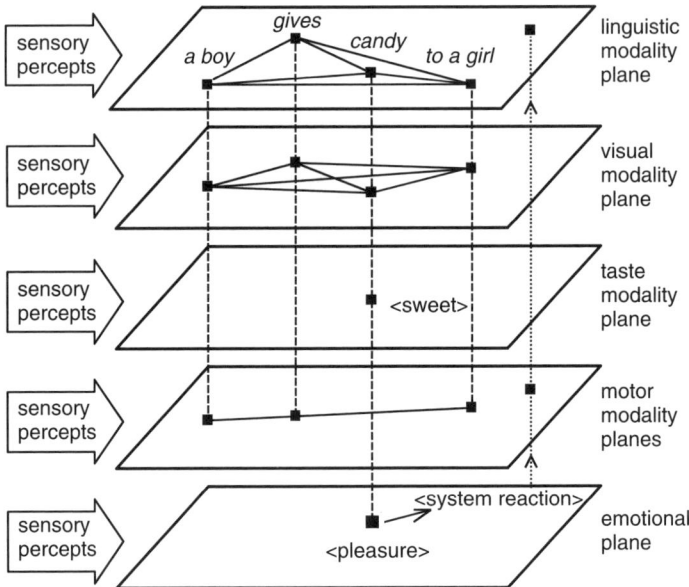

Figure 9.2 The multimodal model of language

on the other planes will evoke linguistic representations of the same kind on the linguistic plane. In this way sentences will evoke multimodal assemblies of related situational representations that allow the paraphrasing of the original sentence.

In Figure 9.2 an example sentence 'A boy gives candy to a girl' is depicted. This sentence is represented on the linguistic modality plane as the equivalent of heard words. On the visual modality plane the sentence is represented as the equivalent of a seen scene. This mental imagery may, however, be extremely simplified; instead of photo-quality images only vague gestalts may be present, their relative position being more important. The action 'gives' will be represented on the visual modality plane, but also on planes that are related to motor actions. The object 'candy' will be represented on the visual plane and taste modality plane and it will also evoke an emotional value response on the emotional evaluation plane. In this way a given sentence should ideally evoke the same percepts that would be generated by the actually perceived situation and, for the other way around, the situation should evoke the corresponding sentence. In practice it suffices that rather vague mental representations are evoked, a kind of model that contains the relationships between the entities.

According to this multimodal model the usage and understanding of language is a process that involves interaction between the various sensory modalities and, at least initially and occasionally, also interaction between the system and the external world.

Words have meanings. The process that associates words with meanings is called grounding of meaning. In the multimodal model of language, word meaning is grounded horizontally to other words and sentences within the linguistic plane and vertically to sensory percepts and attention direction operations. Initially words are acquired by vertical grounding.

9.4.2 Vertical grounding of word meaning

The vertical grounding process associates sensory information with words. This works both ways: sensory percepts will evoke the associated word and the word will evoke the associated sensory percepts, or at least some features of the original percept. The associations are learned via correlative Hebbian learning (see Chapter 3) and are permanent.

Vertical grounding of meaning is based on ostension, the pinpointing of the percept to be associated with the word. Typical entities that can be named in this way are objects (a pen, a car, etc.), properties (colour, size shape, etc.), relation (over, under, left of, etc.), action (writes, runs, etc.), feelings (hungry, tired, angry, etc.), a situation, etc. Certain words are not grounded to actual objects but to the act of attention focus and shift (this, that, but, etc.) and are associated with, for instance, gaze direction change. Negation and affirmation words (yes, no, etc.) are grounded to match and mismatch signals. It should be noted that the vertical grounding process operates with all sensory modalities, not only with vision. Thus the pinpointing also involves the attentive selection of the intended sensory modality and the selection of the intended percept among the percepts of the chosen modality. This relates to the overall attention management in the cognitive system.

The principle of the vertical grounding process with the associative neuron groups is illustrated by the simple example in Figure 9.3. For the sake of clarity words are represented here by single signals, one dedicated signal per word.

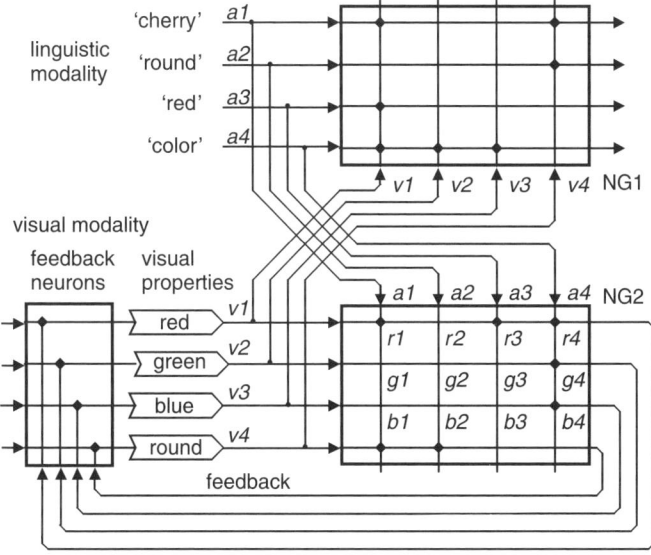

Figure 9.3 A simple example of vertical grounding of a word meaning

Figure 9.3 depicts one associative neuron group $NG1$ of the linguistic (auditory) modality and the simplified perception/response loop of the visual modality with the associative neuron group $NG2$. The neuron group $NG1$ is an improved associator, either Hamming, enhanced Hamming or enhanced simple binary associator, that does not suffer from subset interference. The neuron group $NG2$ can be a simple binary associator.

In this example it is assumed that the visual modality detects the colour features <red>, <green> and <blue> and the shape feature <round>. The names for these properties are taught by correlative learning by presenting different objects that share the property to be named. In this way, for instance, the associative connection between the word 'red' and the visual property <red> is established in the neuron groups $NG1$ and $NG2$. This connection is created by the synaptic weight values of 1 at the cross-point of the 'red' word line and the $a1$ line at the neuron group $NG1$ as well as at the cross-point of the <red> property line and the $b2$ line at the neuron group $NG2$. Likewise, the word 'round' is associated with the corresponding <round> property signal. The word 'colour' is associated with each <red>, <green> and <blue> property signal.

A cherry is detected here as the combination of the properties <red> and <round> and accordingly the word 'cherry' is associated with the combination of these. Thus it can be seen that the word 'cherry' will evoke the properties <red> and <round> at the neuron group $NG2$. The evoked signals are returned into percept signals via the feedback and these percept signals are broadcast back to the neuron group $NG1$ where they can evoke the corresponding word 'cherry'.

Vertical grounding also allows the deduction by 'inner imagery'. For example, here the question 'Cherry colour?' would activate the input lines $b1$ and $b4$ at the neuron group $NG2$. It is now assumed that these activations were available simultaneously for a while due to some short-term memory mechanism that is not shown in Figure 9.3. Therefore the property <red> would be activated by two synaptic weights and the properties <green>, <blue> and <round> by one synaptic weight. Thus the property <red> would be selected by the output threshold circuit. The evoked <red> signal would be returned into the <red> percept signal via the feedback and this signal would be broadcast to the $a1$ input of the neuron group NG1. This in turn would cause the evocation of the signals for the words 'red' and 'colour'. ('Cherry' would not be evoked as the property signal <round> would not be present and subset interference does not exist here.) Thus the words 'red colour' would be the system's answer to the question 'Cherry colour?' Likewise, the question 'Cherry shape?' would return 'round shape' if the word 'shape' were included in the system.

In Figure 9.3 the meaning of words as grounded to visually perceived objects. According to the multimodal model of language words can be vertically grounded to all sensory percepts including those that originate from the system itself. The principle of multimodal grounding is presented in Figure 9.4.

Figure 9.4 shows how auditory percepts are associatively cross-connected to the visual, haptic and taste modalities. An internal grounding signal source is also depicted. These signals could signify pain, pleasure, match, mismatch, novelty,

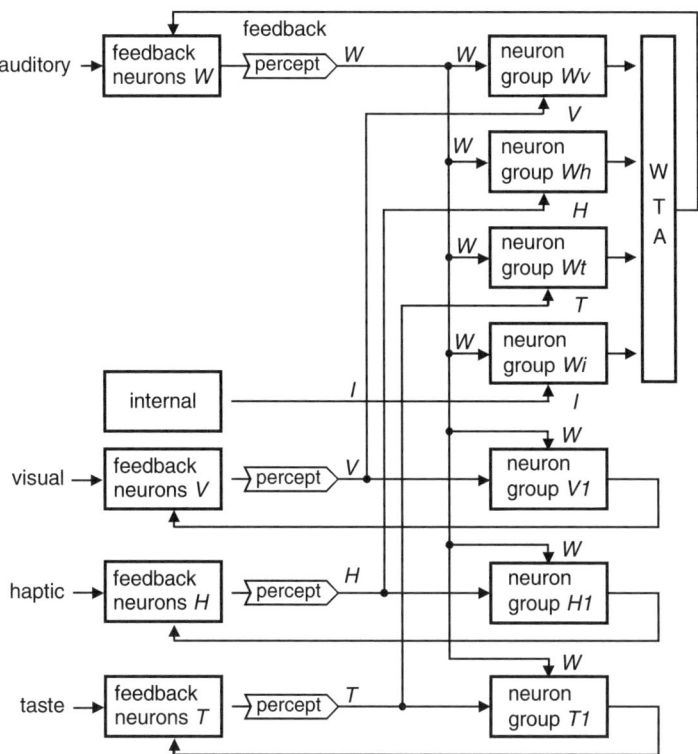

Figure 9.4 Multimodal vertical grounding of a word meaning

etc., conditions. Each modality has its own auditory modality neuron group, which associates percepts from that modality with words so that later on such percepts can evoke the associated words in the neuron groups Wv, Wh, Wt and Wi. Likewise, words are associated with percepts within each sensory modality, the neuron groups $V1$, $H1$ and $T1$. These neuron groups allow the evocation of percepts by the associated words. There is no neuron group that would allow the evocation of the sensations of pain or pleasure. The word 'pain' should not and does not evoke an actual system reaction of pain.

During association there may be several sensory percepts available simultaneously for a given word, which should be associated only with one of the percepts. Two mechanisms are available for the creation of the correct association, namely the correlative Hebbian learning and, in addition to that, the attentional selection. The attended percept should have a higher intensity than the nonattended ones and this higher intensity should be used to allow learning only at the corresponding word neuron group. The resulting associations should be permanent.

The machine may generate rudimentary sentences and even converse by using vertical grounding only. Assume that a small vocabulary has been taught to the machine by association. Thereafter the machine will be able to describe verbally

168 NATURAL LANGUAGE IN ROBOT BRAINS

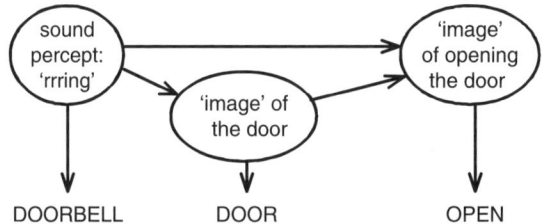

Figure 9.5 The evocation of a vertically grounded sentence

simple situations and the mental ideas that the situation evokes. A simple example is given in Figure 9.5.

In the situation of Figure 9.5 a doorbell rings. The corresponding sound percept evokes the word 'doorbell'. The sound percept evokes also some inner imagery, for instance the 'image' of the door and consequently the 'image' of the opening of the door. (These 'images' are not real images, instead they may consist of some relevant feature signals.) This sequence leads to the sentence 'Doorbell... door... open'. This may be rather moronic, but nevertheless can describe the robot's perception of the situation.

The sensory modalities provide a continuous stream of percepts. This gives rise to the selection problem: how to associate only the desired percept with a given word and avoid false associations. For instance, there may be a need to associate the percept <sweet> with the corresponding word 'sweet' and at the same time avoid the association of the word 'sweet' with any percepts from the visual and haptic modalities. Obviously this problem calls for mechanisms of attention; only attended and emotionally significant signals may be associated, while other associations are prevented. Possible attention mechanisms are novelty detection and emotional significance evaluation for each modality. These may control the learning process at the sensory modality neuron groups and at the modality-specific neuron groups at the auditory modality (neuron group learning control; see Figure 3.10).

9.4.3 Horizontal grounding; syntactic sentence comprehension

Horizontal grounding involves the registration of the relationships between the words within a sentence. This is also related to the sentence comprehension as it allows answers to be given to the questions about the sentence.

Consider a simple sentence 'Cats eat mice'. The simplest relationship between the three words of the sentence would be the association of each word with the two remaining words. This could be easily done with an associative neuron group by assigning one signal to each of the words, as in Figure 9.6.

In Figure 9.6 a dot represents the synaptic weight value 1. Here the output is evoked when word signals are inserted into the vertical lines. The horizontal word signal line with the highest evocation sum will be activated. Thus, it can be deducted that the question 'who *eat mice*' would evoke '*cats*' as the response. (The word 'who' has no relevance here, as it is not encoded in the neuron group.) The question

THE MULTIMODAL MODEL OF LANGUAGE   169

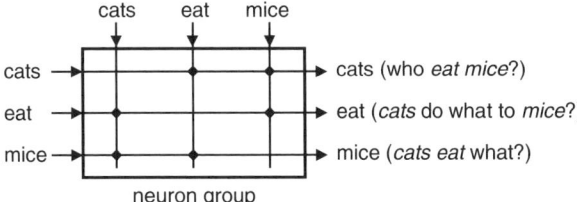

Figure 9.6 Using an associative neuron group to register word-to-word relationships

'*cats* do what to *mice*' would evoke '*eat*' as the response and the question '*cats eat* what' would evoke '*mice*' as the response. (It should be noted that without any vertical grounding the system does not know what the words 'cats', 'eat', 'mice' actually depict.)

So far so good, but what would be the answer to the question 'what do *mice eat*'? The neuron group will output the word 'cats'. However, normally mice do not eat cats and obviously the example sentence did not intend to claim that. In this case the necessary additional information about the relationships of the words was conveyed by the word order. Here the simple associative neuron group ignores the word order information and consequently '*mice eat cats*'–type failures (subject–object confusion) will result. However, the word order information can be easily encoded in an associative neuron group system as depicted in Figure 9.7.

The associative network of Figure 9.7 utilizes the additional information that is provided by the word order and the categorical meaning of the words. (The categorical meaning is actually acquired via vertical grounding.) This is achieved by the use of the Accept-and-Hold circuits AH1, AH2 and AH3. The Accept-and-Hold circuits AH1 and AH2 are set to accept nouns; AH1 captures the first encountered noun and AH2 captures the second noun. The AH3 circuit captures the verb. After the Accept-and-Hold operations the first and second nouns and the verb are available simultaneously. The associative network makes the associations as indicated in Figure 9.7 when a sentence is learned. When the question '*Who eat mice*' is entered the word 'who' will be captured by AH1, 'eat' will be captured by AH3 and 'mice' will be captured by AH2. The word 'who' is not associated with anything, its only function here is to fill the AH1 circuit so that the word 'mice'

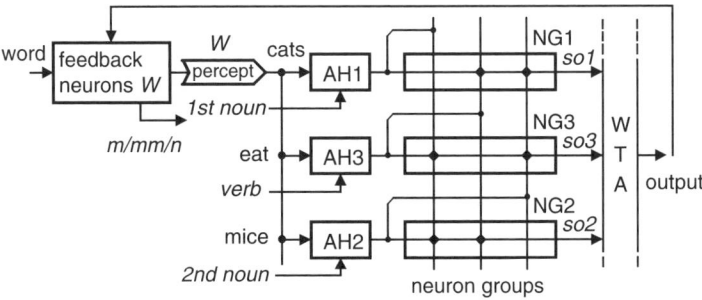

Figure 9.7 Word order encoding in an associative network

will settle correctly at AH2. For this purpose the word 'who' must be defined here as a noun. Now it can be deduced that the question *'Who eat mice'* will evoke 'cats' as the response. Likewise, the question *'Cats eat what'* will evoke 'mice' as the response. The question *'Mice eat what'* will not evoke 'cats' as the response, as 'mice' as the first noun is not associated with 'cats' as the second noun. Thus the system operates correctly here.

The example sentence 'cats eat mice' contains only the subject, verb and object. Next, a more complicated example sentence is considered: 'Angry Tom hits lazy Paul'. Here both the subject (Tom) and the object (Paul) have adjectives. The associative network must now be augmented to accommodate the additional words (Figure 9.8).

The operation of the associative network is in principle similar to the operation of the network in Figure 9.7. The subject–object action is captured by the neuron groups $NG1$, $NG2$ and $NG3$, which all share a common WTA output circuit. Each neuron group has its own input Accept-and-Hold circuit, AH1, AH2 and AH3.

The subject–object action is captured from the incoming sentence by the neuron groups $NG1$, $NG2$ and $NG3$ and the related Accept-and-Hold circuits AH1, AH2 and AH3. The first noun and the second noun Accept-and-Hold circuits AH1 and AH2 are connected. They accept nouns sequentially; the AH1 circuit accepts and captures the first noun and the AH2 circuit captures the second noun. The AH3 circuit accepts the verb.

When the network learns the information content of a sentence it forms associative connections between the words of that sentence. These connections operate via the learned synaptic weights, as indicated in Figure 9.8. The sentence, however, is not stored anywhere in the network.

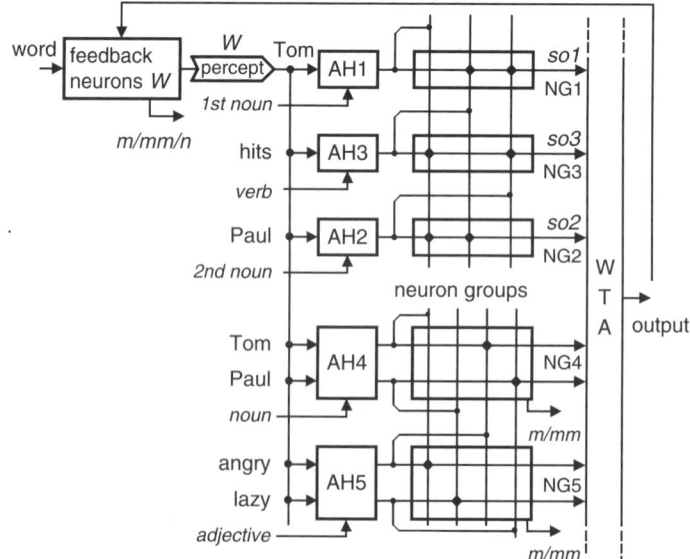

Figure 9.8 The network for the sentence 'Angry Tom hits lazy Paul'

If the network has properly captured the information content of the sentence, it will be able to answer questions about the situation that is described by that sentence. When, for instance, the question '*Who hits Paul*' is entered, the word 'who' is captured by AH1, forcing the word 'Paul' to be captured by AH2. The verb 'hits' is captured by AH3. The associative connections will give the correct response 'Tom'. The question '*Paul hits whom*' will not evoke incorrect responses, as 'Paul' will be captured by AH1 and in that position does not have any associative connections.

The associative neuron groups *NG*4 and *NG*5 associate nouns with their adjacent adjectives. Thus 'Tom' is associated with the adjective 'angry' and 'Paul' with 'lazy'. This is done in the run. As soon as 'Tom' is associated with 'angry', the Accept-and-Hold circuits AH4 and AH5 must clear and be ready to accept new adjective–noun pairs. After successful association the question 'Who is *lazy*' will evoke the response 'Paul' and the question 'Who is *angry*' will evoke the response 'Tom'.

Interesting things happen when the question 'Is *Tom lazy*' is entered. The word 'Tom' will evoke the adjective 'angry' at the output of *NG*5 while the word 'lazy' will evoke the word 'Paul' at the output of *NG*4. Both neuron groups *NG*4 and *NG*5 now have a mismatch condition; the associatively evoked output does not match the input. The generated match/mismatch signals may be associated with words like 'yes' and 'no' and thus the system may be made to answer 'No' to the question 'Is *Tom lazy*' and 'Yes' to the question 'Is *Tom angry*'.

This example has been simulated by a Visual Basic program written by the author. The visual interface of this program is shown in Figure 9.9.

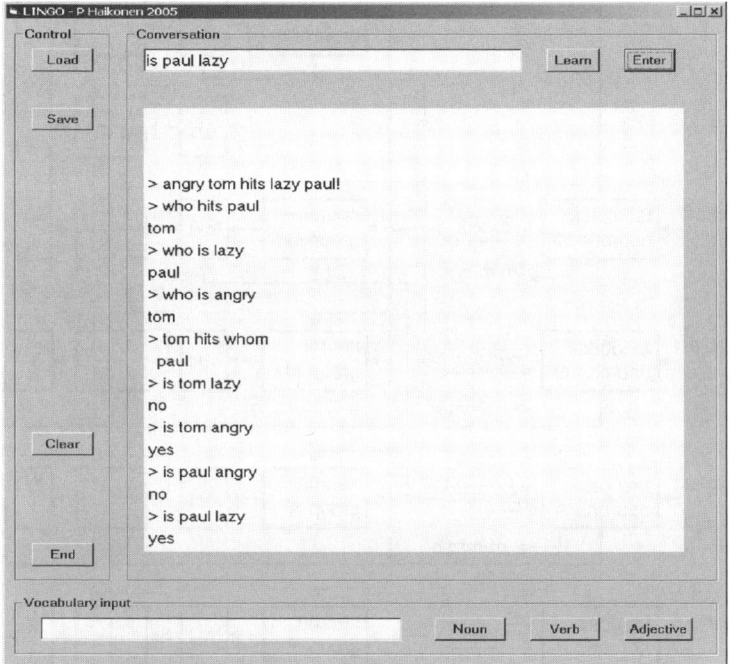

Figure 9.9 Sentence understanding with associative neural architecture, a Visual Basic program

9.4.4 Combined horizontal and vertical grounding

The purpose of the combination of horizontal and vertical grounding is to provide the system with the ability to produce and understand complete sentences. The vertical grounding process alone can produce strings of words that are able to evoke the corresponding sensory percepts and vice versa sensory percepts can evoke the corresponding words. On the other hand, the horizontal grounding process can process word–word associations. However, in the horizontal grounding the real understanding will remain missing, because the meanings of the words are not grounded to anywhere and consequently the process cannot bind sentences to real word occurrences. Therefore, a full language utilization capacity calls for the combination of the horizontal and vertical grounding processes.

The horizontal and vertical grounding processes can be combined by providing cross-associative paths between the linguistic (auditory) modality and the other sensory modalities. Figure 9.10 gives a simplified example, which combines the principles of Figures 9.4 and 9.7.

Figure 9.10 depicts a circuit that combines the processes of horizontal and vertical grounding. The word perception/response loop has the required associative neuron groups for the horizontal word–word connections as before. In addition to these the circuit has the neuron groups Wn and Wv for the vertical grounding of word

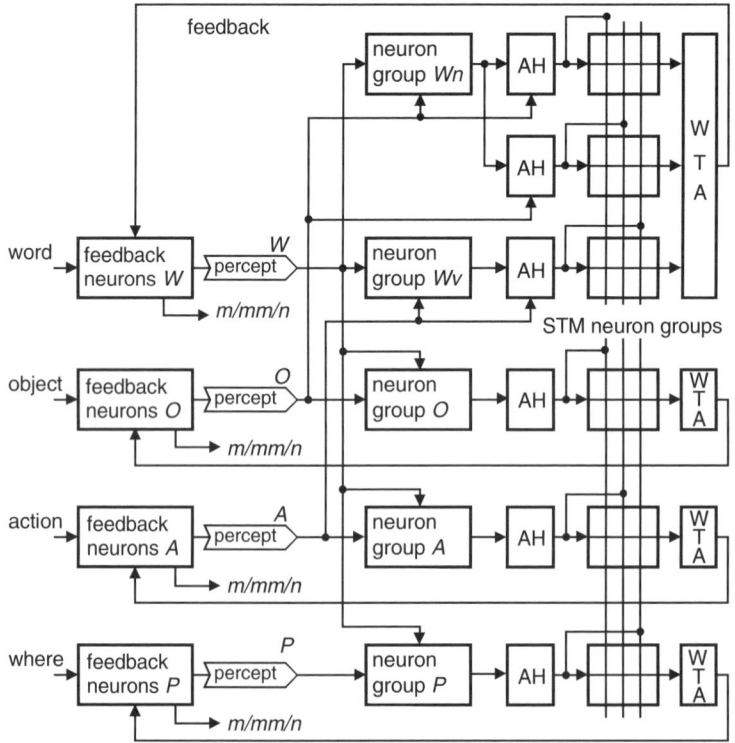

Figure 9.10 Combined horizontal and vertical grounding

meaning. The associative inputs of these neuron groups are connected to the visual object percepts and action percepts. These connections allow the association of a given word with the corresponding percept so that later on this percept may evoke the corresponding word, as described earlier.

The word percept W is broadcast to the visual 'object', 'action' and 'where' perception/response loop neuron groups O, A and P where they are associated via correlative learning with the corresponding object, action and location percepts. Thereafter a given word can evoke the percept of the corresponding entity (or some of the main features of it). The Accept-and-Hold (AH) circuits will hold the percepts for a while, allowing the cross-connection of them in the following short-term memory associative neuron group. This neuron group will maintain a situation model of the given sentence, as will be described in the following.

Figure 9.10 depicts the vertical grounding of word meaning to some of the possible visual percepts only and should be taken as a simplified illustrative example. In actual systems the vertical grounding would consist of a larger number of cross-connections and would be extended to all the other sensory modalities, such as haptic, olfactory, proprioception, etc.

9.4.5 Situation models

The cognitive system perceives the world via its sensors and creates a number of inner representations, percepts, about the situation. These percepts enable a number of associative connections to be made between themselves and memories and learned background information, as seen in Chapter 7, 'Machine Cognition'. The percepts evoke a number of models that are matched against each other and the combination of the matching models constitutes the system's running model of the world. This model is constantly compared to the sensory information about the world and match/mismatch conditions are generated accordingly. If the model and the external world match then the system has 'understood' its situation.

In this framework, language understanding is not different. The actual sensory percepts of the world are replaced with linguistic descriptions. These descriptions should evoke virtual percepts of the described situation and the related associative connections, the situation model, just like the actual sensory percepts would do. This model could then be inspected and emotionally evaluated as if it were actually generated by sensory perception. This is also the contemporary psychology view; language is seen as a set of instructions on how to construct a mental representation of the described situation (Zwaan and Radvansky, 1998; Zwaan, 2004; Zwaan et al., 2004; Zwaan and Taylor, 2006).

The construction of mental representations calls for the naming of the building blocks and assembly instructions, words and syntax. Hearing or reading a story involves the construction of a mental model of the story so far. The understanding of subsequent sentences may involve the inspection of the created mental model in a proper order. This inspection would involve the utilization of inner attention in the form of the 'virtual gaze direction'. Thus, sentences must describe situations, but in addition to that, they must also indicate how the mental models are to be inspected and construed; therefore the meaning of words should also relate to attention guidance. This conclusion is rather similar to that of Marchetti (2006), who proposes that words and language pilot attention; they convey attentional instructions. It is easy to find examples of attention guiding words that indicate the relative position and change of position (e.g. from, to, over, under, left, right, next, etc.). However, words focus attention more generally. The naming of an object focuses attention on that object and its associations. Certain words and word combinations indicate how attention should be shifted and refocused.

In the multimodal model of language the system remembers the situation model of a read story, not the actual text as strings of words. The situation model can be used to paraphrase and summarize what has been read.

In the multimodal model of language situation models arise naturally. Words activate corresponding inner representations via the vertical grounding process. For instance, the word 'book' may evoke some visual features of books. However, this is not all. The cognitive system may have some background information about books, for example that they can be opened, they can be read, etc. The activation of the features of a book will also enable associative paths to this background information, which can thus be evoked depending on the overall context. Likewise, a linguistic sentence will activate a set of representations and these in turn will enable associative paths to a larger set of background information. These evoked representations and their associative connections to background information constitute here the situation model. Thus, in the multimodal model of language the situation model is an imagined situation with the effect of context and background information, evoked by a linguistic description.

A situation model may include: (a) actors, objects and their properties; (b) spatial locations, such as who, what is where; (c) relative spatial locations, such as in front of, above, below, to the right, etc.; (d) action, motion, change, (e) temporal order, such as what was before, what came next, (f) multimodality, such as sensory percepts, motor actions.

As an example the sentence 'Tom gives sweets to Mary' and its corresponding situation model is depicted in Figure 9.11. The words and their order in the sentence 'Tom gives sweets to Mary' evoke percepts in the sensory perception/response loops as follows. 'Tom' evokes a visual feature vector T at the visual object modality for the visual entity $<T>$ and associates this with an arbitrary location P1. The locations correspond to virtual gaze directions and may be assigned from left to right unless something else is indicated in the sentence. 'Gives' evokes a motion percept that has the direction from left to right. 'Sweets' evoke a percept of an object $<S>$, and also a percept at the taste modality. 'Mary' evokes the percepts of the object $<M>$ for which the location P2 will be given. This process creates an internal scene that may remain active for a while after the word–sentence percepts have expired.

This internal scene of the situation model may be inspected by virtual gaze scanning. For instance, the gaze direction towards the left, the virtual location P1,

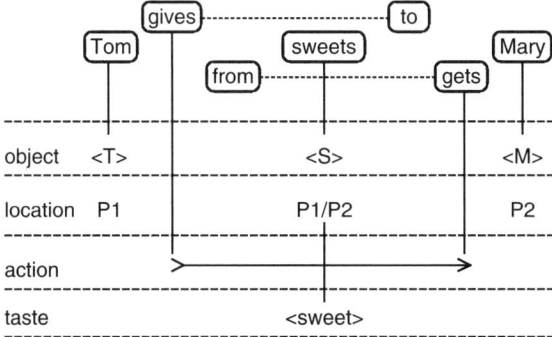

Figure 9.11 A sentence and its situation model

evokes the object $<T>$, which in turn evokes the word 'Tom'. Thus, if the scene is scanned from left to right, the original sentence may be reconstructed. However, if the internal scene is scanned from right to left, paraphrasing occurs. The action percept will evoke 'gets' instead of 'gives' and the constructed sentence will be 'Mary gets sweets from Tom'.

A situation model also includes history. The system must be able to reflect back and recall what happened before the present situation. This operation utilizes short-term and long-term memories and the recall can be executed via associative cues that evoke representations of the past situation.

9.4.6 Pronouns in situation models

In speech pronouns are frequently used instead of the actual noun. For instance, in the following sentence 'This is a book; it is red' the words '*this*' and '*it*' are pronouns. Here the pronoun '*this*' is a demonstrative pronoun that focuses the attention on the intended object and the pronoun '*it*' is a subjective pronoun, which is here used instead of the word 'book'. The pronoun '*it*' allows the association of the book and the quality 'red' with each other.

In associative processing the fundamental difference between nouns and pronouns is that a noun can be permanently associated with the percept of the named object, while a pronoun cannot have a permanent association with the object that it refers to at that moment. 'It' can refer to anything and everything and consequently would be associated with every possible item and concept if permanent associations were allowed. Thus the presentation of the word 'it' would lead to the undesired evocation of 'images' of every possible object. Moreover, the purpose of the sentence 'it is red' is not to associate the quality 'red' with the pronoun 'it' but with the entity that the pronoun 'it' refers to at that moment, the 'book'.

In a situation model the operation of the pronouns like 'it' can be achieved if the pronoun is set to designate an imaginary location for the given object. In this way the pronoun will have a permanent association with the location, which will also be temporarily associated with the intended object. (The imaginary location

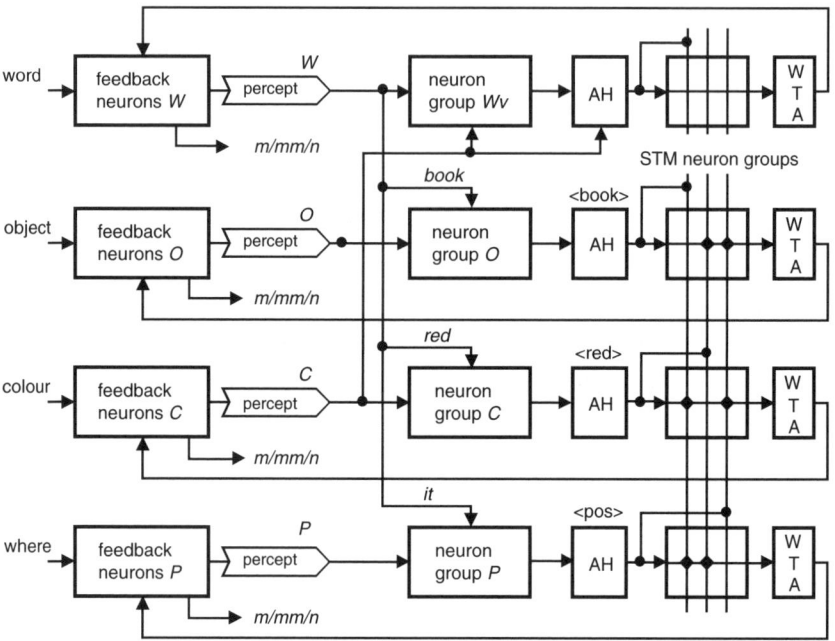

Figure 9.12 Processing the pronoun 'it' via position

would correspond to a location that can be designated by the virtual gaze direction; a 'default' location would be used instead of the many possibilities of the gaze direction.) The processing of the example sentences is depicted in Figure 9.12.

The sentence 'This is a book' associates the object <book> with the position P. The sentence 'it is red' evokes the colour <red> by the word 'red' and the position P by the word 'it'. The position P evokes the object <book>, which is routed via feedback into an object percept and will be subsequently captured by the Accept-and-Hold circuit. At that moment the Accept-and-Hold circuits hold the object <book> and the colour <red> simultaneously and these will then be associated with each other. Thus the 'it' reference has executed its intended act.

9.5 INNER SPEECH

In a system that implements the multimodal model of language the generation of linguistic expressions becomes automatic once the system has accumulated a vocabulary for entities and relationships. The sensory percepts and imagined ones will necessarily evoke corresponding linguistic expressions: speech. This speech does not have to overt and loud, instead it can be silent inner speech. Nevertheless, the speech is returned into a flow of auditory percepts via the internal feedback. Thus this inner speech will affect the subsequent states of the system. It will modify the running inner model and will also be emotionally evaluated, directly and via the modifications to the inner model.

INNER SPEECH   177

Functional inner speech necessitates some amendments to the circuits that are presented so far. The linguistic module is situated within the auditory perception/response module and should therefore deal with temporally continuous sequences. Yet in the previous treatment the words are considered as temporally frozen signal vectors, which operate as discrete symbols. Actually words are temporal signals, which can be represented with reasonable accuracy by sequences of sound features. Associative processing necessitates the simultaneous presence of the sequential sound features, and so the serial-to-parallel operation is required. On the other hand, speech acquisition, the learning and imitation of words, is serial, and so is also the inner silent speech and overt spoken speech. Therefore the linguistic process must contain circuits for both serial and parallel processing.

The basic auditory perception/response feedback loop is serial and sequential and is able to predict sequences of sound feature vectors. The audio synthesizer loop is also serial and is able to learn and reproduce sound patterns. The linguistic neuron groups that operate in the parallel mode must be fitted to these in a simple way. One such architecture is presented in Figure 9.13.

The system of Figure 9.13 is actually the system of Figure 9.1 augmented by the parallel neuron groups for linguistic processing. The sequences of sound feature vectors are transformed into a parallel form by the S/P circuit. Thereafter the operation of the neuron groups $W1$ and $W2$ and the AH circuits is similar to what has been previously described. These circuits output their response word in the parallel form. This form cannot be directly returned as a feedback signal vector to the audio feedback neurons as these can only handle serial sound feature

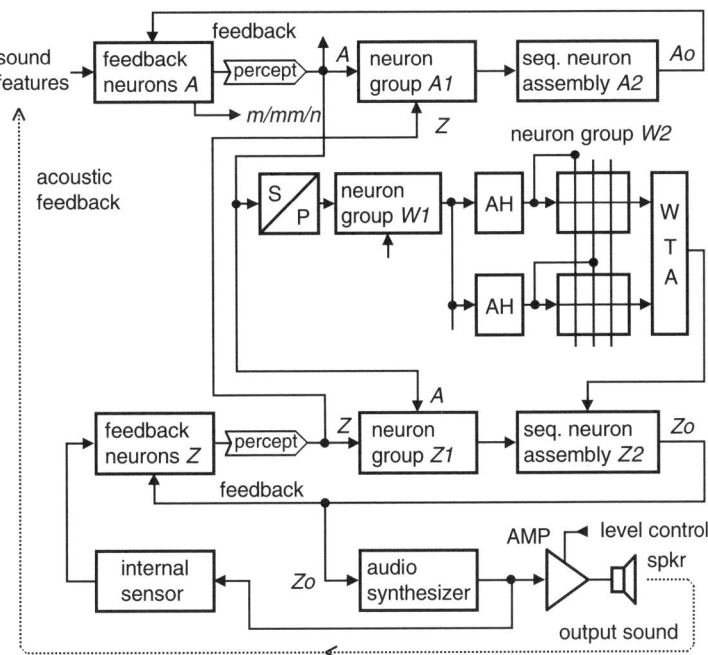

Figure 9.13 Architecture for serial speech

vectors. Likewise, the parallel output words cannot evoke directly any spoken words. Therefore, the output words or possibly syllables are first associated with the corresponding audio synthesizer control vector Zo sequences at the sequence neuron assembly $Z2$. Thereafter a parallel word or syllable representation will evoke the corresponding Zo vector sequences, which then command the synthesis of the corresponding sounds. The timing of the produced word is determined by the audio synthesizer loop sequence neuron assembly $Z2$. When the temporal sequence for a word has been completed then attention must be shifted to another percept, which then leads to the evocation of another word.

The audio synthesizer output is sensed by the internal sensor. The output of this sensor appears as the synthesizer command percept vector Z. This vector is broadcast to the auditory neuron group $A1$ where it is able to evoke corresponding sound feature sequences. Thus the internally generated speech will be perceived as heard speech, as also when the audio output amplifier is disabled. This is the mechanism for silent inner speech. Initially the necessary associations at the neuron group $A1$ are created when the system outputs random sounds and these are coupled back to the auditory perception module via external acoustic feedback.

How can it be guaranteed that the emerging inner speech is absolutely logical and coherent and does not stray from the current topic? There is no guarantee. The more the system learns the more there will be possibilities and the less the inner speech will be predictable. However, this also seems to be a weakness of the human mind, but also its creative strength.

10
A cognitive architecture for robot brains

10.1 THE REQUIREMENTS FOR COGNITIVE ARCHITECTURES

In artificial intelligence cognitive architectures are computer programs with an overall structure that tries to emulate and combine various cognitive functions in order to generate human-like cognition and general intelligence. An artificial cognitive system may try to emulate the human way of cognition, but it does not necessarily try to emulate the exact organization and structure of the human brain. These architectures may consist of separate program modules for the various cognitive functions; there may be individual modules for perception, memory, attention, reasoning, executive control and even consciousness. These architectures describe a certain organization and order of computation between these modules. They may be criticized for not introducing any real qualitative difference, as the executed computations are still those of the conventional computer. It can be argued that a computer program will not readily possess consciousness even if one of its program modules is labelled so.

Thus, the author proposes that a true cognitive architecture would not be a computer program. It would be an embodied perceptive system that would be characterized by the effects that it seeks to produce.

The foremost effect that a true cognitive architecture should produce would be the direct, lucid and enactive perception of the environment and the system itself. The system should perceive the world directly and apparently and not via some indirect sensory data. This effect should be accompanied by seamless sensorimotor coordination, the immediate readiness to execute actions on the entities of the environment.

The lucid perception of the world should be accompanied by processing with meaning. The architecture should support the association of percepts with additional entities and, as a consequence, the creation of associative networks of meaning and models. This would also involve the ability to learn.

A true cognitive architecture should be able to generate both reactive and deliberate responses. The latter would call for the ability to plan, imagine and judge the actions. This, in turn, leads to the requirement of the flow of mental content, world models and abstractions that are detached from the direct sensory percepts. This kind of content could be described as 'an inner world' or 'inner life'.

Humans have the flow of inner speech. An advanced cognitive architecture should also be able to support natural language and inner speech. True cognitive architectures should be designed to support autonomous self-motivated systems such as robots.

10.2 THE HAIKONEN ARCHITECTURE FOR ROBOT BRAINS

A cognitive architecture that integrates the circuits, subsystems and principles presented in the previous chapters is outlined in the following. This architecture seeks to implement the above general requirements for true cognitive architectures.

The cognitive architecture is designed for an embodied perceptive and interactive system, such as a cognitive robot, with physical body, sensors, effectors and system reactions. It is based on the cross-connected perception/response feedback loops and the architecture contains a perception/response feedback loop module for each sensory modality. These modules work concurrently on their own, but they broadcast their percepts to each other and may also cooperate with some or all of the other modules. The architecture supports the processes of multimodal perception, prediction, distributed attention, emotional soundtrack, inner models, imagination, inner imagery, inner speech, learning, short-term and long-term memories, system reactions, machine emotions, good/bad value systems and motivation.

The block diagram depicting the Haikonen cognitive architecture is presented in Figure 10.1, where seven perception/response feedback loop modules are depicted. This is not a fixed specification; other sensory modalities may be added as desired. These additional sensory modalities may even include some that humans do not have, such as the sensing of electric, magnetic and electromagnetic fields, etc.

The structure of these modules follows the principles that are described earlier. Especially blocks of the 'neuron groups' depict the AH-STM/LTM style of circuitry of Figure 7.3, with additional circuits as discussed in previous chapters. The modules communicate with each other via the broadcast lines.

The cognitive system is motivated by the modules 1 and 2. The module 1 evaluates the emotional good/bad significance of sensory and imaginary percepts and triggers suitable system reactions. This module also provides the emotional soundtrack, and facilitates emotional learning. Hard contact or mechanical damage sensors provide pain information. This module is important for the survival of the robot. Here the principles of Chapter 8, 'Machine Emotions', are used.

The physical requirements of the system are detected by the module 2. For this purpose the module may include sensors for energy levels, motor drive levels, mechanical tension, balance, temperature, moisture, etc. These sensors may also provide pain information for the module 1. This module is also responsible for sensing mechanical and electrical soundness of the system. Detected anomalies evoke specific short-term and long-term goals and goal-oriented behaviour.

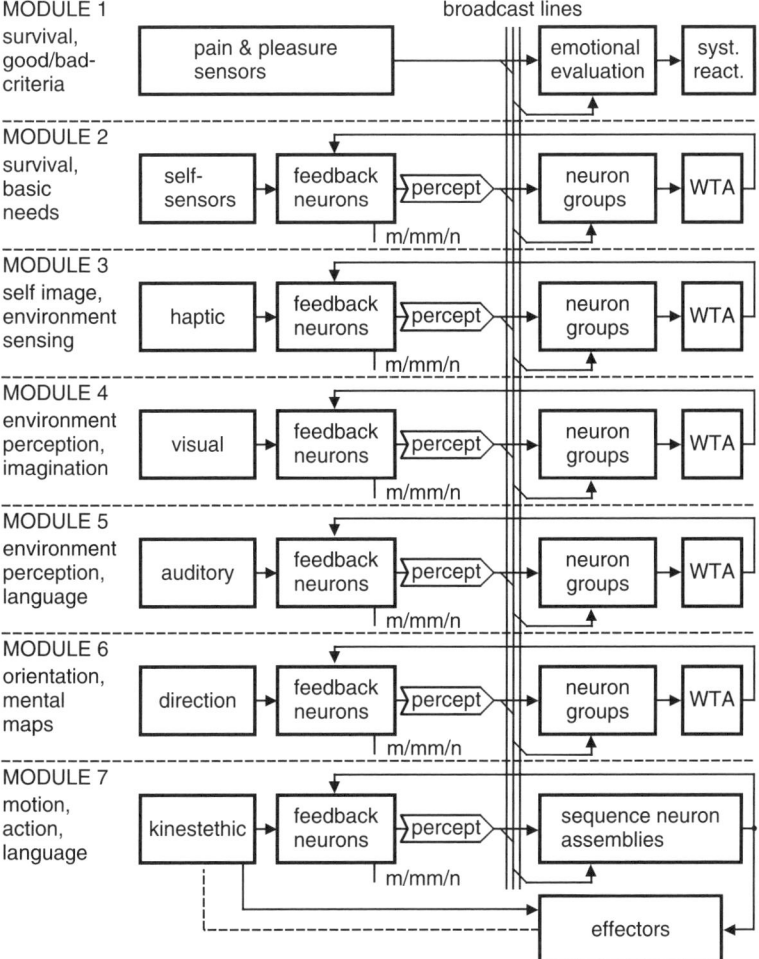

Figure 10.1 The Haikonen cognitive architecture for robot brains

The module 3 is responsible for the sensing of soft contact or touch information. This module can sense the shape and surface of objects. It can also create an inner model of the robot body by touching every reachable point and associating this with kinesthetic location information. The principles of section 5.6, 'Haptic perception', in Chapter 5 are used here.

The module 4 is responsible for the sensing of visual information. This module also facilitates visual introspection, imagination and the use of visual symbols and visual models of objects. The principles of Section 5.7, 'Visual perception', are used here.

The module 5 is responsible for the sensing of auditory information and auditory introspection. This module also facilitates the use of linguistic symbols, words and inner speech. The principles of Section 5.8, 'Auditory perception', and Chapter 9, 'Natural Language in Robot Brains', are used here.

The module 6 provides 'absolute' and relative direction information about the robot's orientation. This information is utilized in the generation of temporary and long-term inner maps of surroundings. The principles of Section 5.9, 'Direction sensing' in Chapter 5 are used here.

The module 7 is responsible for the learning and generation of motion and motor actions. The operation of this module is closely related to the haptic, visual and auditory sensory modules. The principles of Chapter 6, 'Motor Actions for Robots', are used here.

In this kind of cross-connected architecture each module may be trying to broadcast its percepts to all the other modules. However, it is not feasible that every module would receive and accept broadcasts from every other module at the same time, as each of these broadcasts might try to evoke completely different and potentially clashing percepts and responses in the receiving modules. Therefore an internal attention mechanism is needed. This mechanism should select the most pertinent broadcasts at each moment that the modules should accept and receive. The inner attention should also work the other way around; a module that processes an important stream of percepts should be able to request a response from other modules. Both these attention control situations may be realized by signal intensity modulation and threshold control by the means that have been already discussed. The response request may be realized by the principle of Figure 5.5. Emotional significance may be used in attention control by the principles of Figure 8.1.

Does this architecture conform to the requirements for true cognitive architectures? The foremost effect that the architecture should produce is the direct, lucid and enactive perception of the environment and the system itself. How would this architecture produce the internal 'lucid appearance' that objects and entities are located out there? For humans things are out there because they intuitively know what it would take in motor action terms to reach out for them. Humans also notice that when they turn, the relative directions towards the things change, but the things remain stationary and do not move along their movements. The enabling fact behind the possibility to perceive things to be out there is that the real origination point of the sensation, the sensor, is not taken as the origination point of the sensed stimuli. Thus the retina is not taken as the location of images and the eardrums are not taken as the origination points of the sounds. This architecture reproduces these conditions. In this architecture sensory information is represented in a way that allows the association of a sensation with an external point of origin, one that is coherent with the results of explorative motor actions, such as turning the head. This is enabled by the associative coupling between the sensory modalities and the motor modalities.

How would this architecture process meaning? This architecture is based on the perception/response feedback loops that inherently operate with meaning. The intrinsic meanings of the signals are grounded to the feature detectors, but each signal and group of signals may convey additional associated meanings and may operate as a symbol. These meanings are learned.

The perception/response feedback loops also allow match and mismatch detection between predicted or desired conditions and actual conditions. This facilitates the adjustment of behaviour to suit the situation and the system's needs. In the introspective mode the perception/response feedback loops disengage from the sensory perception and generate percepts of imagined events. In this way the system may try actions without actually executing them as described in Chapter 7, 'Machine Cognition'.

10.3 ON HARDWARE REQUIREMENTS

The Haikonen architecture utilizes the associative neuron group as the basic building block and the perception/response feedback loop as the next-level assembly. At this moment these are not freely available as integrated circuits and the question is: 'If one were to design such chips then how many neurons should be integrated?' The human brain has some 10^{11} neurons and some 10^{14} synapses. On the other hand, the honey bee has only about 960 000 neurons (Giurfa, 2003) and yet has impressive cognitive and behavioural capacities. The required capacity may also be estimated from the sensory requirements. Auditory perception might do with around 100 feature signals and visual perception might generate some 1000 feature signals. In a neuron group each signal would correspond to a neuron and the number of synapses for each neuron would depend on the number of cross-couplings. If each neuron were coupled to every other neuron, then the number of synapses for each neuron would be the total number of neurons. In small systems this might be the case; in large systems all neurons are not connected to every other neuron. Thus, a small 100 neuron group might have 10 000 synapses and a 1000 neuron group could have one million synapses. A complete system would, of course, utilize several neuron groups. Systems with minimal sensory capacity and clever preprocessing could do with a rather small number of neurons and synapses. However, the integration of a large number of synapses should not be a problem now that humans are able to produce low-cost gigabyte memories (around 10^{10} one bit memory locations).

Another problem with the hardware realization is the large number of interconnecting wires. Within an integrated neuron group the wiring reduces into a simple geometry and should be manageable. The wiring between chips, however, becomes easily impractical due to the immensely large number of parallel lines. This can be solved by serial communication between chips. A simple protocol that allows the cross-connection of chips in a transparent way can be easily devised. The serial communication can be very fast and will not deteriorate the overall speed performance of the system. However, the serial communication will turn the system into a temporally sampling system.

11
Machine consciousness

11.1 CONSCIOUSNESS IN THE MACHINE

The ultimate goal of machine cognition research is to develop autonomous machines, robots and systems that know and understand what they are doing, and are able to plan, adjust and optimize their behaviour in relation to their given tasks in changing environments. A system that succeeds here will most probably appear as a conscious entity. Here the issues of the apparent properties of the human consciousness and those of the proposed machines are discussed.

At first, it is useful to make the difference between the concepts of *the contents of consciousness* (such as the immediate percepts and memories) and *the mechanism* that makes the system aware of the contents of consciousness. The machine consciousness research is mainly interested in the mechanism of consciousness, while the contents of the machine's consciousness, as interesting as that will be, are here a secondary issue. Additional confusion may arise from the fact that the phrases 'to be conscious of something' and 'to be aware of something' are often also used in the meaning 'to know' or 'to have the knowledge of something'.

A conscious agent is aware of its existence in a meaningfully perceived world. Therefore, in order to understand the world the agent must deal with meanings. This is a fundamental requirement. If a machine that is claimed to be conscious does not process meanings then the machine is not conscious. Real-world meanings can be extracted by suitable sensors and perception processes. In humans these processes produce a direct, lucid appearance of the world, without the perception of any material-carrying media such as the actual neural signals. This situation should be reproduced in a conscious machine.

11.1.1 The immateriality of mind

The first principle of the author's model for a conscious machine is the creation of a lucid, apparently immaterial perception of the sensed entities. Humans perceive thoughts and percepts as immaterial, referring directly to the object entities. This

---

Robot Brains: Circuits and Systems for Conscious Machines Pentti O. Haikonen
© 2007 John Wiley & Sons, Ltd

relates to the philosophical mind–body problem: the mind is immaterial while the body and its actions are not. How can an immaterial mind control the material body? For the other way around, how can a material brain cause an immaterial mind? The following solution is proposed here: the mind is not really immaterial, it only appears to be so, because humans cannot perceive the neurons and neural processes that carry thoughts and percepts. However, this approach leads to the consequential question: 'How can a material mechanism carry information and remain unnoticed?' As the solution the author proposes transparent systems and processes (Haikonen, 2003a). A transparent system operates only with the information content, while the material mechanism is not observed and does not enter into the logic of the information content.

The perception process utilizes feature signals with their meanings grounded to the properties of the outside world. The transparency of these signals can be achieved by modulation principles; the signals are taken as carriers and the information as the modulation. Circuits that operate only on the modulation do not 'see' the carrier, the actual material basis of the representation. Examples of transparent systems include radio, television, telephone and sound reproduction systems. In each of these systems the qualities of the reproduced content are perceived, not the carrier waves or the transistors or other components that facilitate the operation. In a cognitive machine the transparent process can be created by 'rigid' causal connections between the representations and the represented features. This connection gives to each signal an intrinsic meaning and this is what the system 'sees', not the material carrier. The circuitry and signals of the machine remain outside of the carried information. Thus, for the machine the information is immaterial.

This principle also applies to the mental content of the machine. The mental content is made available via introspection, which is realized via the feedback loops that return the results of the inner processes back to the perception points. In this way the introspected content will appear in terms of sensory percepts and their grounded meanings. For instance, the inner speech is perceived as 'heard' speech, although not necessarily with all the acoustic qualities of real speech.

11.1.2 The reportability aspect of consciousness

The second principle of the author's model for a conscious machine relates to attention and reportability. Consciousness requires objects; if humans are conscious, they are conscious of something. Consciousness and perception are related; every entity that humans are aware of is a percept. This applies also to mental content, inner speech and imagery. These also appear to humans in terms of sensory percepts, as 'heard' speech, 'seen' images, 'felt' system states. Should both sensory perception and introspective perception cease then also consciousness would vanish, as there would be nothing to be conscious of. However, it is known that nonconscious perception and action is also possible. Therefore perception would seem to be a necessary but not sufficient condition for consciousness. The required additional functionality can be identified by the comparison of conscious and nonconscious percepts and acts.

Nonconscious and conscious acts are different. Nonconscious acts are automatic and are executed without overall attention and memory making, and consequently the details of the actions cannot be reported or recalled. Nonconscious acts may be performed fluently and with great precision while conscious acts may involve deliberation about the details of the act to be executed; all mental units focus their attention on the act. This in turn may make the execution slower and less fluent. For instance, swallowing is normally a nonconscious act and no notice is taken of it. However, when a pill needs to be swallowed, a person may become very conscious of the act, even to the extent that swallowing becomes all but impossible. Obviously it is possible to remember and be able to report this incident for a while.

On the other hand, while driving to work a dangerous crossing may be negotiated without any conscious awareness of the event, because attention has been paid to other thoughts. Afterwards there might be wonder at how it was done as there was not the faintest memory of the incident. Obviously, a complicated feat had been executed without overall awareness, like a zombie (if these exist). If the event could be remembered and if a report of it had been made, then it could be said that it was done with awareness.

Examples like these seem to show that the difference between nonconscious and conscious acts is the absence or presence of overall attention and the resulting reportability.

This aspect of consciousness can be readily understood within the framework of machine architecture. The machine consists of a large number of perception/response loop modules. Each module operates basically in the same way whether the overall operation is 'conscious' or 'nonconscious'. Every moment each module broadcasts its percepts to every other module. All percepts are not accepted, instead they have to compete with each other at every module's associative inputs. Also, if the receiving module is very busy, it will not accept any broadcasts. A successful broadcast will evoke related percepts, either sensory or memory, in the receiving module. These percepts are of the sensory kind of receiving module. In this way the broadcasting module 'recruits' the receiving module to handle the broadcasting module's topic and, in return, may readily accept the response broadcast from the recruited module. As a result closed associative loops may arise. The machine becomes 'conscious' about a topic when the various modules cooperate in this way in unison and focus their attention on the percepts about this topic. This will involve a wealth of cross-connections and subsequently the forming of associative memories. Therefore the 'conscious' event can be remembered by the machine for a while and can be reported in terms of the various modalities such as sounds, words, gestures, drawings, written text, etc. It can be seen that this kind of 'consciousness' is expensive, as the recruited modules cannot process any other topic at the same time. Thus only one thing can be processed at a time; while in the 'unconscious' mode the modules could process many things in parallel. This would also seem to apply to humans; for instance it is difficult to hear when one is reading an interesting book.

11.1.3 Consciousness as internal interaction

An operational mode, which would seem to correspond to the reportability aspect of consciousness, arises in a multimodal perceptual system from the associative interconnections between the modalities. According to the 'multimodal model of language' the vertically and horizontally grounded inner speech is one manifestation of these interconnections. If these interconnections break down, the associative evocation process fails and the flow of inner speech and imagery, the 'stream of the contents of consciousness', ceases. Indeed, in the human brain the situation seems to be similar. Von der Malsburg (1997, 2002) has proposed that humans experience different degrees of consciousness, where the difference is made by the degree of communication between parts of the brain. Von der Malsburg uses the word 'coherence' for this state of successful communication. There is also some experimental proof: Massimi *et al.* (2005) have noticed that during dreamless sleep, when there is no consciousness (or inner speech), neural communication between the different parts of the cerebral cortex breaks down while local activities may still exist.

Baars' global workspace model (Baars, 1997, 2003) and Dennett's multiple drafts model (Dennett, 1991) also depend on the interaction between the different parts and modules of the system. The author's model has some similarities with these models.

Baars' global workspace model assumes a large number of nonconscious specialized autonomous networks that compete for access to the so-called global workspace area. The winner will be able to 'post' its message to this workspace and this message is then broadcast to each of the nonconscious processes. The broadcast part of the contents of the global workspace is seen as the focus of attention and the contents of consciousness at that moment. Baars' global workspace model is a 'theatre' model, where the global workspace is a kind of stage where the contents of consciousness bathe in the spotlight of attention. Baars compares the autonomous networks to a decentralized audience that is sitting in the darkened theatre (Baars, 2003). In Figure 11.1 the autonomous network B wins the competition and sends its message to the global workspace. This message is then 'in the spotlight of attention' and will be broadcast to the other autonomous networks.

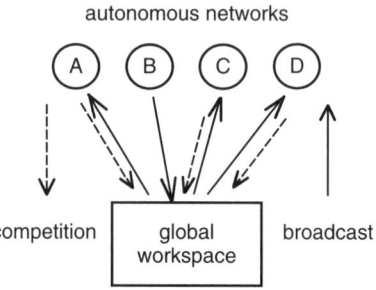

Figure 11.1 The principle of Baars' Global workspace model

The author's model involves a large number of autonomous perception/response modules. Each module tries to broadcast its percept to the other modules. The winning percept will be accepted by all or most of the other modules and this percept will then associatively evoke representations of the receiving module's own kind. These in turn will be accepted by the winning module and closed associative loops will be formed; the modules will operate in unison on the winning topic. In other words, the attention of each module is focused on the winning topic. This topic is seen as the contents of consciousness at that moment.

In Figure 11.2 the module B succeeds in broadcasting its percepts to the other modules. These will respond by their own percepts, either sensory or introspective, which relate to the percepts of the module B. In this way all the modules will be paying attention to the same topic, thus allowing the formation of cross-associations and memory of the event and the other hallmarks of a conscious moment, as discussed before.

It should be noted that in the author's model the modules communicate directly with each other; therefore it is also possible that some modules engage in mutual communication while other modules are doing something else. In humans this kind of action may be manifested, for instance, in 'the bed-time story reading effect' (Haikonen, 2003a, p. 252). It is possible to read aloud a story without any conscious attention. One may think about completely different matters while reading and will not be able to report afterwards what the story was about. Yet the visual and speech motor modules must have been communicating with each other. A similar effect may sometimes take place when one is driving home or to work. One may suddenly notice that no memory exists about how a familiar but dangerous crossing was negotiated.

In Baars' model the global workspace can be understood as the site of consciousness. In the author's model there is no specific site for consciousness. There is no box that could be labelled as 'consciousness'; all neuron groups operate in the same way whether the overall action is conscious or not.

It can be seen that both Baars' model and the author's model propose that the contents of consciousness are determined by the focus of global attention. However, the author's model is not a theatre model. It does not utilize a global workspace, which is seen to be redundant as the modules can communicate and compete directly with each other. There are also other important differences. Baars does not explain

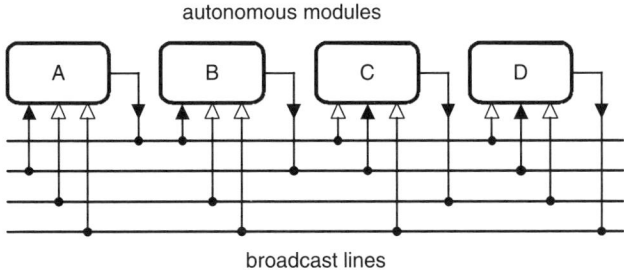

Figure 11.2 The principle of Haikonen's conscious architecture model

the apparent immateriality of percepts and the mental content and does not specify how inner speech could arise, while the Haikonen model is specific on both issues.

In view of the above the author argues that consciousness is not an executing agent; it does not do anything and it does not have causal powers. It is not even a product of evolution. Instead, consciousness is the way that the operation appears to the transparent cognitive system when the interaction between the different parts and modules of the system takes place.

11.2 MACHINE PERCEPTION AND QUALIA

Consciousness involves the direct and lucid perception of the world and its entities. These percepts appear to have different qualities (qualia): the redness of red, the coolness of ice, the wetness of water, etc. Obviously these qualities are related to the qualities of the perceived entities; redness is related to the spectral properties of the reflected light from an object, coolness is related to the temperature of an object and so on. Human senses are not perfect and sometimes the subjective quality (quale) does not match the real quality of the sensed object, for instance a very cold object may give rise to a burning sensation.

Qualia differentiate the various percepts from each other. The auditory, visual, haptic, pain, pleasure, etc., percepts all have different qualities. It is difficult to imagine how the percepts could be separated from each other if there were no qualia. In that case all sounds and sights would be similar, no colours could be seen and no grey scale shades could be distinguished. How could anything be separated from anything? However, in the brain information is carried by mutually similar neural signals. These signals are not red, cold or wet, yet they manage to carry this information. In the consciousness philosophy the question of qualia has a central position. How can mutually similar neural signals carry differentiable varieties of quality-related information and how can they cause a different subjective feel of these qualities? This is the so-called hard problem of consciousness (Chalmers, 1995).

The first part of the qualia question is easy to solve in engineering terms. In the outlined cognitive machine the percept signal vectors that describe the outside world would contain all necessary information as such, and will differentiate percepts from each other directly due to the hardwired intrinsic meanings of the individual signals. This situation is not very different from, for instance, colour television, where colour information about the three primary colours is carried by three mutually similar signals. The intended colour is causally encoded in the circuit path (actual circuit or radio way). The electric colour signals do not possess the property of any colour.

The second part of the qualia question is more obscure. If the percepts were differentiated from each other by the circuitry, then would any additional subjective feel of qualia in the machine be necessary at all? alternately, would this differentiation appear as some kind of qualia as a trivial consequence of the grounding of the meaning of signals? Due to the transparency principle the machine will not perceive signals and signal vectors; only the carried information can cause effects. Therefore, would the system perceive and report this information as having some kind of qualia? The situation might be similar in the human brain, so it may be asked, and philosophers have done so: 'Are subjective qualia real and necessary or might they only be some kind of illusory by-products of the perception process (Dennett, 1991)?'

Technically it would suffice that in the machine the different percepts from the different sensory modalities are differentiated from each other by hardwired architecture and/or the organization of information processing. In this way signals and vectors become to have a fixed intrinsic meaning that is causally connected to the corresponding quality of the external world via the corresponding feature detectors. The sensed quality would be carried as the (binary on/off) modulation on the signal in the dedicated signal line. No further consideration of qualia or the design of some 'exotic qualia processes' is necessary. The machine will be able to report 'red' and 'wet' when it senses these.

11.3 MACHINE SELF-CONSCIOUSNESS

11.3.1 The self as the body

As stated before, consciousness requires objects: if I am conscious, I am conscious of something. Self-consciousness relates to states where the object of awareness is an aspect of the system self. A cognitive robot is an entity with a body and a 'brain' with mental content, the 'thoughts'. A self-aware entity must be aware of its body and thoughts and recognize these as its own. The ownership calls for an owner, a fixation point that can be associated with the owned properties.

The body is the natural owner and the most concrete fixation point for the self. For example, Damasio (2000, p. 22) has proposed that the sense of self is grounded to representations (percepts) of the body. The percepts of the body and body functions could provide self-related mental entities that could be associated with other aspects of the self. In this way associative nets of self-information could arise. Thus, the build-up of a robot's self-consciousness would begin with the awareness of the robot's mechanical body and its states, and this could then be developed into the sense of ownership of the robot's mental content.

Obviously it is rather easy to provide the robot with an array of sensors that deliver information about its mechanical body and the body's physical states. Also, the proposed perception/response loop allows introspection of the internal mental states in terms of external percepts. Thus, basically, the proposed system should be able to create the required associations for self-consciousness.

A self-conscious entity must be able to make a difference between its bodily self and the environment. It must know which physical objects are parts of its own body and which ones are parts of the environment, which action and change is due to external causes and which ones are self-generated. The robot acquires information about these via perception; therefore it must have a mechanism that

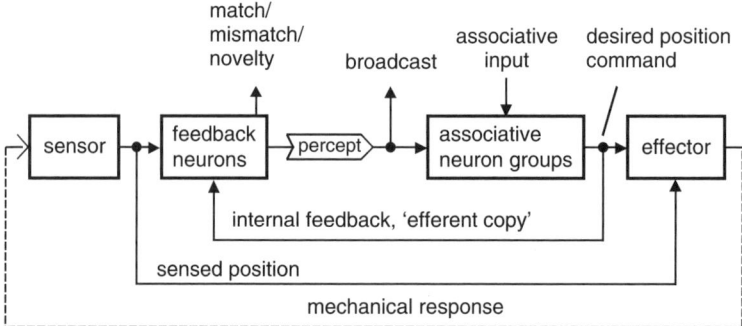

Figure 11.3 The perception/response loop for motor action

can indicate which percepts relate to the outside environment and which ones are self-related. Figure 11.3 depicts how this can be accomplished in the motor perception/response loop.

In Figure 11.3 a desired position for a moving part is evoked via the associative input. The effector begins to execute the necessary motion and will drive the moving part until the sensed position information from the sensor equals the desired position. Self-consciousness calls for the awareness that the system itself is executing the action, as it might well be that the action could take place due to external forces. This information is facilitated by the internal feedback from the output of the associative neuron groups to the feedback neurons. The feedback neuron group detects the match/mismatch/novelty condition between the sensed information and the feedback information. A match condition would indicate that a desired position exists and that it matches the sensed position; the goal has been generated by the system itself. A mismatch condition would indicate that a desired goal exists, but it has not been reached. A novelty condition would indicate that no internal desired position exists; if there is movement it is due to external causes.

In neuroscience a rather similar mechanism for the recognition of self-caused actions has been proposed. The brain signals that correspond to the internal feedback are called 'efference', 'efferent copy' or 'corollary discharge' (Gallagher, 2000; Tsakiris and Haggard, 2005).

11.3.2 The experiencing self

Self-consciousness involves more than the mere observation of bodily and mental things; it involves the concept of the experiencing self. My body is moving – I am walking. My body has an upright posture – I am standing. In these cases 'my body' is the 'I' that is doing something. This is easy enough, but how about 'the experiencing self'? I feel pleasure, I feel pain; who is feeling it if not I, me, myself? The 'I' seems to refer here to 'the self', but exactly what would be this feeling entity?

Would the system be the feeling entity? The system is in a state of pleasure, state of pain, nobody is feeling it. To feel the pain is to be in pain. Here it is possible to be deceived by verbal expressions. The way that the system states are described leads to the thought that there must be a special discrete entity, a self, which is perceiving, feeling and experiencing the percepts and system states. There is no such requirement. The cognitive system is able to function without this kind of additional self-entity. The concept of the 'experiencing self' is a linguistic illusion. It arises from the way in which the system is describing its inner states. 'I am feeling pain' is not to be taken literally, word by word; instead the whole sentence is a label for a system state. There is no discrete 'feeling self'; there is only the system.

Would this kind of concept of self arise in a machine? Abstract concepts arise from concrete ones. In this case naming would help. A name could be associated with the robot, for instance, by inputting the name-word in one way or other while patting the robot (provided that the robot has touch sensors). Thereafter the name will evoke the bodily touch sensations and the robot will 'know' that the name is related to the robot itself. Consequently, the name will catch the robot's attention.

11.3.3 Inner speech and consciousness

Inner speech constitutes one form of reports that are required for the reportability aspect of consciousness. However, inner speech may not be a necessary condition for consciousness as there may be other means of reporting. Nevertheless, the presence of inner speech is an indication of the presence of the internal interaction aspect of consciousness, as without the cross-module communication no meaningfully grounded inner speech can exist. The author has argued that the inner speech provides a part of the stream of consciousness (Haikonen, 2006). Inner speech is also one folk psychology manifestation of thinking and consciousness.

In humans inner speech seems to amplify self-consciousness. In inner speech it is possible to engage in thoughts about thoughts: 'I am thinking now' and in doing so am aware of having thoughts. Inner speech has been seen as a tool for introspection and one of the most important cognitive processes involved in the acquisition of information about the self and the creation of self-awareness (Morin and Everett, 1990; Morin, 1993, 2005; Siegrist, 1995; Schneider, 2002; Steels, 2003; Haikonen, 2003a, pp. 256–260).

With inner speech one can comment on one's own situation. Morin (2005) sees this self-talk as a device that can lead to social self-awareness. (The author has argued elsewhere that basic self-awareness does not require social interaction; see the 'Hammer test' in Haikonen, 2003a, p. 161.) As a part of social interactions humans are subject to comments about themselves, the way they are and behave. Self-talk also allows them to imitate internally the act of appraisal; they may echo the patterns of others' comments directly or as first-person transformations. Humans may ask themselves: 'Why did you do this stupid thing?' or 'Why did I do this stupid thing?' Originally it was your mother that posed the question (Haikonen, 2003a, p. 240). In this way inner speech turns into a tool for self-evaluation, which in turn will affect our self-image: who we are, what we want.

These effects are not excluded in an advanced cognitive machine by any universal principle.

11.3.4 The continuum of the existence of the self

Self-awareness also includes the sense of the continuum of existence: what one has been, what one has done, what one is doing now and what would be the expected future. This sense is also related to the perception of passing time, what is happening now and what happened a moment ago. However, this sense is more than that, as is demonstrated by the recovery of this sense after sleep. When we wake up we know who we are, where we are and what we are supposed to do. A cognitive robot should experience its existence in a similar way, and also when it recovers from possible periods of power-saving inactivity. The sense of passing time can be created in the robot by the principles of Section 7.4, 'The perception of time', in Chapter 7.

The body represents a continuum of existence and all things that are associated with the body will share this continuum and will be evoked when the awareness of the body is restored after a time of inactivity. In this way the continuum of existence can be created in the robot.

11.4 CONSCIOUS MACHINES AND FREE WILL

Humans are supposed to possess 'free will'; they can choose what to want, what to do and what to think. Some researchers attribute 'free will' to consciousness; others see that willed actions arise from subconscious processes. Nevertheless, 'free will' would seem to be an integral property of mind. Therefore, should the conscious machine also possess it, and how could it be implemented?

'Free will' is demonstrated in the decisions that are made. It appears that humans can choose between given alternatives as they please. In this process 'free will' can be seen as the opposite of causality. If the decision process could be causally tracked then the decision would be deterministic; it would be the result of cause and consequence chains without any real freedom of choice. Thus, strict causality would seem to exclude the existence of 'free will'. Conventional machines are causal systems and cannot support any noncausal 'free will'. Should noncausal processes therefore be included in the cognitive machine, for example in the form of random noise generators? On the other hand, total randomness can hardly be equated to 'free will' as this would again exclude one's own ability to affect the outcome of the decision process. The flipping of a coin does not lead to 'free will' decisions. Therefore, is it reasonable at all to try to find a point between deterministic causality and total randomness for the home of 'free will'? I do not think so.

If 'free will' is so important then why do humans try so hard to do without it? We try to find rational reasons for our decisions; for instance, we try to find the best car for our requirements, we try to find the cutest woman for our girlfriend and wife, and so on. In this way humans are actually replacing any possible 'free will' with rational criteria.

'Free will' decisions are not really random; in fact they may be very predictable. Friends may know very well what we want and what kind of 'free will' decisions we are likely to make. We may decide to do something due to external social pressures, because we feel that it is expected from us. We may also do something if it gives us emotional satisfaction. Sometimes we may even decide to do something useful that conforms to our long-term goals and values. These are rather clear causal reasons for deciding and doing something and the utilization of these does not necessarily involve any 'free will'. However, this may not be a strict deterministic process. The human mind may entertain conflicting interests, short-term pleasure may compete with long-term goals and the final decision may hinge on any additional stimuli, however small and random. Afterwards we may declare that we decided it in this way because we wanted it to be so. Sometimes, however, we may wonder why we ever made that decision.

Thus 'free will' may not be noncausal or random after all. Yet the actor may be able to make decisions that can serve the actor's needs and goals. Perhaps it might be better not to use the philosophically controversial term 'free will' at all. Instead the concept 'own will' could be used, which might describe much better decision processes that are based on one's own values, goals and yearning for pleasure, etc. These criteria are also present in the cognitive machine that is outlined here; the machine will direct its actual and mental actions by these, and apply them in any branching points calling for decisions (see Section 8.7, 'Machine motivation', in Chapter 8). Therefore the machine will have its 'own will'.

11.5 THE ULTIMATE TEST FOR MACHINE CONSCIOUSNESS

The state of being conscious is a highly subjective experience. We can test our fellow humans by asking them 'Are you conscious now?' The victim of our pursuit may answer 'Yes', but this is not necessarily a proof of consciousness, as the response may be generated by a blind stimulus–response reaction. Likewise we might stick our victim with a needle and see what the reaction would be; a conscious being is supposed to feel pain. Once again, the response might be only a nonconscious reaction. Then there are some unfortunate people who are born without the ability to perceive pain. We do not have any test that could directly show that another person does indeed have a mental experience similar to ours. The only reason that allows some credibility to the consciousness of other human beings is the similarity of our build and biology. As these are the same then perhaps also our subjective experiences should be similar. If we are conscious then we can accept that others are also conscious when they claim so.

Machines are not similar to humans. How then can we test and determine whether a machine would be conscious? It would be useless to ask the machine if it was conscious. After all, it would be very easy to program a computer to respond to questions like that by 'Yes' or even 'Why yes, absolutely'. The mere observation of external behaviour, as in the famous Turing test (Turing, 1950), is not enough because we can easily be fooled. In this context the Turing test is worthless.

One approach towards the determination of consciousness in a machine could be based on the detection of some of the hallmarks of consciousness. For instance, Aleksander and Dunmall (2003) list five 'axioms', functions that are assumed to be typical to a conscious system. These are: (a) the sense of place at the centre of the rest of the world, (b) the faculty of imagination, (c) the ability to direct attention, (d) the ability to plan and (e) emotions in the decision process. The presence of these functions in a machine could be detected in various technical ways. The good point here is that what is sought can be defined exactly. On the other hand, the shortcoming is that presently it is not known for sure whether these functions are sufficient for consciousness or have anything to do with consciousness at all. Nevertheless, it is better to have some criteria instead of none at all, and these criteria are useful when different would-be conscious machines are assessed.

The author has proposed another hallmark-based test for machine consciousness (Haikonen, 2000, 2003a). It is assumed that thinking and consciousness have some distinct properties, like the perceived mental content that in humans appears partly as the flow of inner imagery and inner speech. Inner speech may not be a necessary condition for consciousness, but a system may have to be conscious, at least in some specific sense (e.g. the coherent cooperation way; see Von der Malsburg, 1997; Haikonen, 2003a), in order to have the flow of inner speech *with grounded meanings*. Also, to have inner speech is to have mental content and, after all, a system cannot be aware of its mental content if it does not have any. Therefore the presence of meaningful inner speech should strongly point towards the presence of consciousness.

Thus the consciousness test should show that the machine has the flow of inner speech with grounded meanings and it is aware of it as such and of its content. The existence of the flow of inner speech and imagery can be determined from the architecture of the machine and can also be detected in the actual working hardware. For instance, inner speech can be transformed into text or heard speech via speech synthesis allowing us to actually see and hear what and how the machine is thinking. What remains to be demonstrated is the machine's own awareness of these. There we may have to rely on the machine's own report. The test for machine consciousness would now be that (a) the machine is able to report that it has inner imagery and inner speech and (b) it can describe the contents of these and recognize these as its own product. Also it must be known that the machine produces this mental content meaningfully and not as a preprogrammed string of words and imagery. Therefore on the basis of the construction of the machine it would have to be known that it can have a flow of inner imagery and inner speech and can indeed be able to perceive these as such and can recognize them as its own products, and not as percepts of any external stimuli. It would also have to be known that the machine does have concepts like 'I', 'to have' and 'inner imagery'. Thus here the mere reproduction of blindly learned strings of words like 'I have inner imagery' would not count as a proof. In this way it is possible to monitor and see that the machine thinks and is able to refer to its own thoughts as the owner and originator of these, and also how the thoughts are affected by and refer to the environment. Via these points it can be verified that a thinking machine has been created that is aware of its own thoughts.

Another aspect of consciousness, the direct perception of the world and its entities, can be determined from the working principles and architecture of the machine. A complete machine consciousness test should also determine the existence of the subjective *feel* of qualia and especially the presence of the *subjective feel* of pain and pleasure in the machine. (The function of qualia as the differentiation of signals and the functions of pain and pleasure is supposed to be realized as described before.) This is a tricky issue as there are no such tests for humans yet, so how could these be devised for the machine? The machine may report that it is experiencing real pain. Should this be taken at face value or should it be dismissed? I do not know; perhaps we will have to wait until a machine makes such a statement and see what we can make of it in the light of our instruments.

Perhaps, sometime in the future a consciousness meter might be devised, one that is able to separate the material and mental processes of a man or a cognitive robot and give a meter reading for the depth of the consciousness. At this moment, however, this kind of ultimate test would appear as science fiction, as depicted by this cartoon drawn by my son Pete. This underground-style cartoon is inspired by the philosophical Cartesian mind–body dualism and the difficulty of testing the highly subjective state of consciousness. Here the scientist is indeed able to separate the material and the mental – the answer is found, but perhaps not in the way that was hoped for.

The mad scientist and the ultimate test for machine consciousness

© Pete Haikonen 2006

Cartesian mind–body dualism maintains that consciousness is a property of an immortal soul that eventually goes to heaven (or the other place). Machine consciousness research is not related to this kind of consciousness and does not seek to diminish the value of human beings in any way. Robots will not have immortal souls.

11.6 LEGAL AND MORAL QUESTIONS

The science fiction writer Isaac Asimov anticipated the dangers of autonomous robots and presented three laws for robots that were supposed to prevent any harm to humans (Asimov, 1942). These laws were:

1. A robot may not injure a human being or, through inaction, allow a human being to come to harm.
2. A robot must obey orders given to it by human beings except where such orders would conflict with the first law.
3. A robot must protect its own existence as long as such protection does not conflict with the first or second laws.

These laws appear to be functional at first sight, but may actually lead to unexpected and unfortunate outcomes, as Asimov's stories cleverly show.

A conscious robot will be an autonomous operator and will therefore have the capacity to execute actions on its own initiative. It will make decisions based on the given tasks and goals as well as its training, experience and values. However, even the wisest men may make mistakes that sometimes may lead to extensive damage and even loss of life. Obviously there is no guarantee that a robot would not make serious mistakes ever. When humans cause damage, the liability is clear – a scapegoat will be found. However, what would happen if a conscious robot caused serious damage; who would be liable? Would it be the owner of the robot, the trainer of the robot, the manufacturer of the robot or the designer? Would it be the robot itself? Who should be punished and who should pay for the damage? The author suggests that insurance policies should be developed to cover these cases.

If humans should deem that a robot can be responsible for its actions then what could be the consequences? Would a robot be a legal person? If so, should it then also have some rights? Responsibility means that the actor can understand the consequences of its actions and can tell if these are good or bad. It also means that the actor has a choice, it can choose between alternatives, between being good or bad. Without the possibility and freedom to choose there can be no responsibility. What should be done to a robot that chooses to be bad? Should misbehaving robots be punished? Humans can be punished by causing pain or depriving them of pleasure; punishment involves emotion. Emotionally neutral consequences are hardly punishments or rewards at all and will not adjust the actor's value system in any way. So, how do you punish a machine? Obviously this is related to the question of machine pain, pleasure, emotions and value systems.

What actually is pain? The author has proposed that in the machine pain is a system reaction that disrupts attention and in this way causes the functional consequences of pain. This is by definition and therefore does not really say anything about the feeling of pain. Likewise, machine pleasure is defined here via its functional consequences. This functionality allows the use of pain and pleasure and the expectation of these as motivational factors, and therefore pain and pleasure are also available for reward and punishment. Thus, the misbehaving robot could be punished and in this way made to adjust its value system so that in the future the bad behaviour would not be repeated.

There is a distinct possibility that a cognitive robot would actually *feel* pain. It may even be possible that the robot might become aware of the possibility of the termination of its own existence and consequently begin to fear its own annihilation. This leads to the questions: Will a 'Royal Society for the Prevention of Cruelty to Machines' be needed? Would it be ethically proper and correct to create entities that can genuinely feel pain and abhor their unavoidable end of existence? Perhaps not. But then, we do make babies, don't we?

Here we are getting ahead of things. So far not a single machine has reported plausibly that it is in pain. In fact not a single machine that is able to plausibly *feel* the faintest discomfort has been designed and built. I do not think that it would be useful to design machines (should the principles of this book carry so far) that would be overwhelmed by the feeling of pain or the fear of it, as this kind of machine could hardly do anything useful.

Epilogue

THE DAWN OF REAL MACHINE COGNITION

True machine cognition technology will succeed in areas where the symbolic processing of 'good old-fashioned artificial intelligence' (GOFAI) has so remarkably goofed. If it does not, then we have not understood and done it correctly. After all we do have one working example of a cognitive machine, the brain. Properly executed machine cognition should help us to proceed from the imperfect traditional information technologies to new levels of performance – processing with meaning. This will enable the transition from speech recognition to speech understanding, from pattern recognition to scene understanding, from sentence parsing to story understanding, from statistical 'learning' to cognitive learning and from numerical simulation to free imagination. Conscious autonomous robots will do things that humans cannot safely do; they will go to places where we cannot go. They will have experiences of their own. They will be able to share those experiences with us and in this way enhance the human experience. Perhaps someday a cognitive robot will ask 'Where did I come from?' and wonder that how can it be that those humans who, due to their biological nature, are not related to machines in any way, could possibly have given it a mind that can ponder its own existence.

References

Aleksander, I. (1996). *Impossible Minds My Neurons My Consciousness*, Imperial College Press, London.
Aleksander, I. (2005). *The World in My Mind, My Mind in the World: Key Mechanisms of Consciousness in People, Animals and Machines*, Imprint Academic, UK.
Aleksander, I. and Dunmall, B. (2003). Axioms and tests for the presence of minimal consciousness in agents, in O. Holland (Ed.), *Machine Consciousness*, Imprint Academic, UK, pp. 7–18.
Aleksander, I. and Morton, H. (1993). *Neurons and Symbols*, Chapman and Hall, London.
Ashby, R. (1952). *Design for a Brain*, Chapman and Hall, London.
Asimov, I. (1942). Runaround, *Astounding Science Fiction*, March 1942.
Baars, B. J. (1997). *In the Theater of Consciousness*, Oxford University Press, New York.
Baars, B. J. (2003). The global brainweb: an update on global workspace theory, Guest editorial, *Science and Consciousness Review*, October 2003.
Biederman, I. (1987). Recognition-by-components: a theory of human image understanding, *Psychological Review*, **94**, 115–147.
Bregman, A. S. (1994). *Auditory Scene Analysis: The Perceptual Organization of Sound*, Bradford Books/MIT Press, Cambridge, Massachusetts
Chalmers, D. J. (1995). Facing up to the problem of consciousness, *Journal of Consciousness Studies*, **2**(3), 200–219.
Chrisley, R., Clowes R. and Torrance S. (2005). Next-generation approaches to machine consciousness, in R. Chrisley, R. Clowes and S. Torrance (Eds), *Proceedings of the AISB05 Symposium on Next Generation Approaches to Machine Consciousness: Imagination, Development, Intersubjectivity, and Embodiment*, The Society for the Study of Artificial Intelligence and the Simulation of Behaviour, UK, pp. 1–11.
Churchland, P. S. and Sejnowski, T. J. (1992). *The Computational Brain*, MIT Press, Cambridge Massachusetts.
Damasio, A. R. (2000). *The Feeling of What Happens*, Vintage, UK.
Damasio, A. R. (2003). *Looking for Spinoza: Joy, Sorrow and the Feeling Brain*, Harcourt Inc., Orlando, Florida.
Davis, D. N. (2000). Minds have personalities – emotion is the core, in *Proceedings of the AISB'00 Symposium on How to Design a Functioning Mind*, University of Birmingham, UK, pp. 38–46.
Dennett, D. C. (1991). *Consciousness Explained*, Little, Brown and Company, Boston, Massachusetts.
Dodd, W. and Gutierrez, R. (2005). The role of episodic memory and emotion in a cognitive robot, in *Proceedings of the IEEE International Workshop on Robot and Human Interactive Communication (RO-MAN)*, Nashville, Tennessee, 13–15 august 2005, pp. 692–697.

Duch, W. (1994). Towards artificial minds, in *Proceedings of the 1st National Conference on Neural Networks and Their Applications*, Kule, 12–15 april 1994, pp. 17–28.

Duch, W. (2005). Brain-inspired conscious computing architecture, *The Journal of Mind and Behavior*, **26** (1–2), 1–22.

Eccles, J. C. (1994). *How the Self Controls Its Brain*, Springer-Verlag, Berlin.

Eliasmith, C. (1996). The third contender: a critical examination of the dynamicist theory of cognition, *Philosophical Psychology*, **9**, 441–463; reprinted in P. Thagard (Ed.) (1998), *Mind Readings: Introductory Selections in Cognitive Science*, MIT Press, Cambridge, Massachusetts.

Eliasmith, C. (1997). Computation and dynamical models of mind, *Minds and Machines*, **7**, 531–541.

Franklin, S. (1995). *Artificial Minds*, MIT Press, Cambridge, Massachusetts and London.

Franklin, S. (2003). IDA: a conscious artifact?, in O. Holland (Ed.), *Machine Consciousness*, Imprint Academic, UK, pp. 47–66.

Gelder, T. J. van (1998). The dynamical hypothesis in cognitive science, *Behavioral and Brain Sciences*, **21**, 615–628.

Gelder, T. J. van (1999). Dynamic approaches to cognition, in R. Wilson and F. Keil (Eds), *The MIT Encyclopedia of Cognitive Sciences*, MIT Press, Cambridge, Massachusetts, pp. 243–245.

Gelder, T. van and Port, R. (1995). It's about time: overview of the dynamical approach to cognition, in R. Port and T. van Gelder (Eds), *Mind as Motion: Explorations in the Dynamics of Cognition*, Bradford Books/MIT Press, Cambridge, Massachusetts, pp. 1–43.

Gibson, J. J. (1966). *The Senses Considered as Perceptual Systems*, Houghton Mifflin, Boston, Massachusetts.

Giurfa, M. (2003). The amazing mini-brain: lessons from a honey bee. *Bee World*, **84**(1), 5–18.

Goldstein, E. B. (2002). *Sensation and Perception*, 6th edn, Wadsforth, USA.

Grand, S. (2003). *Growing up with Lucy*, Weidendfield and Nicholson, London.

Gregory, R. (2004). The blind leading the sighted, *Nature*, **430**, 19 August 2004, 836.

Haikonen, P. O. (1999a). Finnish Patent 103304.

Haikonen, P. O. (1999b). *An Artificial Cognitive Neural System Based on a Novel Neuron Structure and a Reentrant Modular Architecture with Implications to Machine Consciousness*, Dissertation for the degree of Doctor of Technology, Series B: Research Reports B4, Applied Electronics Laboratory, Helsinki University of Technology.

Haikonen, P. O. (2000). An artificial mind via cognitive modular neural architecture, in *Proceedings of the AISB'00 Symposium on How to Design a Functioning Mind*, University of Birmingham, pp. 85–92.

Haikonen, P. O. (2002). Emotional significance in machine perception and cognition, in *Proceedings of the Second IASTED International Conference on Artificial Intelligence and Applications*, ACTA Press, Anaheim, California, pp. 12–16.

Haikonen, P. O. (2003a). *The Cognitive Approach to Conscious Machines*, Imprint Academic, UK.

Haikonen, P. O. (2003b). US Patent 66 25588.

Haikonen, P. O. (2005a). Artificial minds and conscious machines, in D. N. Davis (Ed.), *Visions of Mind: Architectures for Cognition and Affect*, Idea Group Inc, USA, pp. 286–306.

Haikonen, P. O. (2005b). You only live twice: imagination in conscious machines, in R. Chrisley, R. Clowes and S. Torrance (Eds), *Proceedings of the AISB05 Symposium on Next Generation Approaches to Machine Consciousness: Imagination, Development, Intersubjectivity, and Embodiment*, The Society for the Study of Artificial Intelligence and the Simulation of Behaviour, UK, pp. 19–25.

Haikonen, P. O. (2006). Towards streams of consciousness; implementing inner speech, in T. Kovacs and J. Marshall (Eds), *Proceedings of the AISB06 Symposium*, Vol. 1,

The Society for the Study of Artificial Intelligence and the Simulation of Behaviour, UK, pp. 144–149.

Hameroff, S. R. (1994). Quantum coherence in microtubules: a neural basis for emergent consciousness?, *Journal of Consciousness Studies*, **1**, 98–118.

Herbert, N. (1993). *Elemental Mind: Human Consciousness and the New Physics*, Dutton, New York.

Hinton, G. E., McClelland, J. L. and Rumelhart, D. E. (1990). Distributed representations, in M. A. Boden (Ed.), *The Philosophy of Artificial Intelligence*, Oxford University Press, New York, pp. 248–280.

Holland, O. (2004). The future of embodied artificial intelligence: machine consciousness?, in F. Iida, R. Pfeifer, L. Steels and Y. Kuniyoshi (Eds), *Embodied Artificial Intelligence*, Springer-Verlag, Berlin, pp. 37–53.

Holland, O. and Goodman, R. (2003). Robots with internal models: a route to machine consciousness?, in O. Holland (Ed.), *Machine Consciousness*, Imprint Academic, UK, pp. 77–109.

Johnson-Laird, P. (1993). *The Computer and the Mind*, 2nd edn, Fontana Press, UK.

Kawamura, K. (2005). Cognitive approach to a human adaptive robot development, in *Proceedings of the IEEE International Workshop on Robot and Human Interactive Communication (RO-MAN)*, Nashville, Tennessee.

Kawamura, K., Dodd, W., Ratanaswasd, P. and Gutierrez, R. (2005). Development of a robot with a sense of self, in *Proceedings of the 6th IEEE International Symposium on Computational Intelligence in Robotics and Automation (CIRA)*, Espoo, Finland, 27–30 June, 2005.

LeDoux, J. (1996). *The Emotional Brain*, Simon and Schuster, New York.

McCulloch, W. and Pitts, W. (1943). A logical calculus of the ideas immanent in nervous activity, *Bulletin of Mathematical Biophysics*, **5**, 115–133.

McGurk, H. and MacDonald, J. W. (1976). Hearing lips and seeing voices, *Nature*, **264**, 746–748.

Marchetti, G. (2006). A presentation of attentional semantics, *Cognitive Processes*, **7**, 163–194.

Marr, D. (1982). *Vision: A Computational Investigation into the Human Representation and Processing of Visual Information*, W. H. Freeman, San Francisco, California.

Massimi, M., Ferrarelli, F., Huber, R., Esser, S. K., Singh, H. and Tononi, G. (2005). Breakdown of cortical effective connectivity during sleep, *Science*, **309**, 30 September 2005, 2228–2232.

Moore, B. C. J. (1973). Frequency difference limens for short-duration tones, *Journal of the Acoustical Society of America*, **54**, 610–619.

Morin, A. (1993). Self-talk and self-awareness: on the nature of the relation, *The Journal of Mind and Behavior*, **14**, 223–234.

Morin, A. (2005). Possible links between self-awareness and inner speech: theoretical background, underlying mechanisms and empirical evidence, *Journal of Consciousness Studies*, **12**(4–5), April–May 2005.

Morin, A. and Everett, J. (1990). Inner speech as a mediator of self-awareness, self-consciousness, and self-knowledge: an hypothesis, *New Ideas in Psychology*, **8**(3), 337–356.

Nairne J. S. (1997). *The Adaptive Mind*, Brooks/Cole Publishing Company, USA.

Newell, A. and Simon, H. A. (1976). Computer science as empirical enquiry: symbols and search, *Communications of the Association for Computing Machinery*, **19**, 113–126.

O'Regan, J. K. and Noë, A. (2001). A sensorimotor account of vision and visual consciousness, *Behavioral and Brain Sciences*, **24**.

Penrose, R. (1989). *The Emperor's New Mind*, Oxford University Press, Oxford.

Plutchik, R. (1980). *Emotion: A Psychoevolutionary Synthesis*, Harper and Row, New York.

Port, R. F. (2000). Dynamical systems hypothesis in cognitive science, Draft entry for amy Lockyer (assoc. Ed.), *Encyclopedia of Cognitive Science*, Macmillan Reference Ltd, London.

Rosenblatt, F. (1958). The perceptron: a probabilistic model for information storage and organization in the brain, Cornell Aeronautical Laboratory, *Psychological Review*, **65**(6), 386–408.

Rosenfield, I. (1995). *The Strange, Familiar and Forgotten*, Picador, UK.

Schachter, S. and Singer, J. E. (1962). Cognitive, social and physiological determinants of emotional state, *Psychological Review*, **69**, 379–399.

Schneider, J. F. (2002). Relations among self-talk, self-consciousness, and self-knowledge, *Psychological Reports*, **91**, 807–812.

Selfridge, O. G. (1959). Pandemonium: a paradigm for learning, in *Symposium on the Mechanization of Thought Processes*, HM Stationary Office, London.

Shannon, C. E. (1948). A mathematical theory of communication, *Bell System Technical Journal*, **27**, July and October, 1948, 379–423 and 623–656.

Siegrist, M. (1995). Inner speech as a cognitive process mediating self-consciousness and inhibiting self-deception, *Psychological Reports*, **76**, 259–265.

Sloman, A. (2000). Introduction: models of models of mind, in *Proceedings of the AISB'00 Symposium on How to Design a Functioning Mind*, University of Birmingham, pp. 1–9.

Steels, L. (2003). Language re-Entrance and the 'inner voice', in O. Holland (Ed.), *Machine Consciousness*, Imprint Academic, UK, pp. 173–185.

Taylor, J. G. (1992). Towards a neural network model of the mind, *Neural Network World*, **6**(92), 797–812.

Taylor, J. G. (1997). Neural networks for consciousness, *Neural Networks*, **10**(7), October 1997, 1207–1225.

Taylor, J. G. (1999). *The Race for Consciousness*, A Bradford Book, MIT Press, London.

Thelen, E. and Smith, L. B. (1994). *A Dynamic Systems Approach to the Development of Cognition and Action*, MIT Press, Cambridge, Massachusetts.

Trehub, A. (1991). *The Cognitive Brain*, MIT Press, London.

Treisman, A. (1998). The perception of features and objects, in R. D. Wright (Ed.) *Visual Attention*, Oxford University Press, New York, pp. 26–54.

Tsakiris, M. and Haggard, P. (2005). Experimenting with the acting self, *Cognitive Neuropsychology*, 2005, **22** (3/4), 387–407.

Turing, A. (1950). Computing machinery and intelligence, *Mind*, **LIX**(2236), October 1950, 433–460.

Valiant, L. G. (1994). *Circuits of the Mind*, Oxford University Press, New York.

Von der Malsburg, C. (1997). The coherence definition of consciousness, in M. Ito, Y. Miyashita and E. T. Rolls (Eds), *Cognition, Computation and Consciousness*, Oxford University Press, Oxford, pp. 193–204.

Von der Malsburg, C. (2002). How are neural signals related to each other and to the world?, *Journal of Consciousness Studies*, **9**(1), January 2002, 47–60.

Wallace, R. (2005). *Consciousness: A Mathematical Treatment of the Neuronal Workspace Model*, Springer-Verlag Inc., New York.

Wiio, O. (1996). *Information and Communication*, Publications Series 1C/1/1996, University of Helsinki Department of Communication.

Zwaan, R. A. (2004). The immersed experiencer: toward an embodied theory of language comprehension, in B. H. Ross (Ed.), *The Psychology of Learning and Motivation*, Vol. 44, Academic Press, New York, pp. 35–62.

Zwaan, R. and Radvansky, G. (1998). Situation models in language comprehension and memory, *Psychological Bulletin*, **123**(2), 162–185.

Zwaan, R. A. and Taylor, L. J. (2006). Seeing, acting, understanding: motor resonance in language comprehension, *Journal of Experimental Psychology: General*, **135**, 1–11.

Zwaan, R. A., Madden, C. J., Yaxley, R. H. and Aveyard, M. E. (2004). Moving words: dynamic mental representations in language comprehension, *Cognitive Science*, **28**, 611–619.

Index

Accept-and-hold circuit   51
Afterimage   96
Aleksander, I.   ix, 196
Anterograde amnesia   143
Apparent motion illusion   95
Artificial Intelligence   1, 201
Artificial neural networks   4
Ashby, R.   4
Asimov, I.   198
Association
  interference in   32
  of vectors   17
Associative
  evocation   17
  function   17
  learning   17
  neural networks   17
  neuron   19
  neuron group   24
  predictor   57
  priming   105
Associator
  bipolar binary   29
  enhanced Hamming distance   30
  enhanced simple binary   31
  Hamming distance   30
  linear   22
  simple binary   26
  with continuous values   28
Attention
  auditory   99
  inner   139, 174, 182
  visual   86
Attention   139
Auditory
  front-back ambiguity   110
  motion detection   111
  perception   98
  scene   98
  spectrum analysis   99

Baars, B. J.   ix, 188, 189
Background information   138, 173, 174
Background knowledge   160
Bed-time story reading effect   189
Biederman, I.   71
Binocular distance estimation   93
Bregman, A. S.   98

Cartesian theater   10
Chalmers, D. J.   190
Change blindness   81
Chrisley, R.   ix
Churchland, P. S.   22, 43
Classification   18, 35, 70
Clowes, R.   ix
Cognition
  computational   3
  dynamical systems   4
  neural network   4
  quantum model   5
Cognitive architecture   6, 179
Cognitive functions   179
Cognitive science   2, 140
Communication   9, 160
  theory   160
  robot-human   156
Comparison, of patterns   80
Connectionism   4
Consciousness   185
  contents of   185
  the hard problem of   190
Context   69, 71, 91, 103, 138, 139, 160, 174
Corollary discharge   96, 192

Damasio, A. R.   149, 191
Davis, D. N.   149
Deduction
　by exclusion   147
　causal   146
Delayed learning   129
Dennett, D. C.   ix, 188, 191
Direction sensing   111
Distributed signal representations   11
Dodd, W.   149
Dreams   144
Duch, W.   ix
Dunmall, B.   196
Dynamical systems   4

Eccles, J. C.   5
Echoic memory   99, 105, 140, 162, 163
Efference   192
Efferent copy   192
Eliasmith   5
Emotional
　decision-making   153
　learning   180
　significance   149, 150
　soundtrack   150, 152, 157, 180
Emotions
　facial expressions of   156
　machine   149
　the system reactions theory of   154
Everett, J.   193
Exclusive-OR interference   37
Exclusive-OR operation   42

Facial expressions   156
Fear   149, 155, 156
Feature detector   72
Feedback
　control   117
　neuron   73
　neuron group   77
　vector   77
Filter bank   99
Folk psychology   2
Formant   102
Fourier transform   99
Fovea   86, 122
Franklin, S.   3
Free will   2, 5, 194, 195
Functions
　computable   42
　noncomputable   42

Gallagher, S.   192
Gaze direction
　control   122
　and fovea   86
　moving towards   129
　and visual attention   86
　and visual memory   87
Gelder van, T. J.   4
Gestalts   85, 90
Gibson, J. J.   69
Giurfa, M.   183
Global workspace   188, 189
Goldstein, E. B.   96
Goodman, R.   5
Grand, S.   5
Grandmother representation   11
Gregory, R.   69
Grounding of meaning   151, 159, 164, 165
Gutierrez, R.   149

Haggard, P.   192
Haikonen associative neuron   19
Haikonen cognitive architecture   180
Hameroff, S. R.   5
Hammer test   193
Hamming distance   30, 42
Haptic perception   82
Herbert, N.   5
Hidden Markow process   71
Hinton, G. E.   11
Holland, O.   5
Homunculus   10

Illusion, apparent motion   95
Illusory contours effect   90
Imagination   79, 145
Immaterial mind   185
Inflection   160
Information theory   9
Inner
　attention   139, 174, 182
　imagery   2, 166, 180, 196
　life   180
　map   182
　models   90, 180
　speech   2, 176, 180, 193, 196
　world   180
　world model   157
Intelligence
　general   1
　specific   1

Inter-aural intensity difference 106
Inter-aural time difference 106
Interference
  exclusive-OR 37
  subset 27, 36
Interference in associators 32
Introspection 79, 181, 186, 191, 193
Introspective perception 76, 79

Johnson-Laird, P. 138

Kawamura, K. 5
Kinesthetic perception 81
Korsakoff's syndrome 143

Language
  in human communication 160
  machine understanding of 159
  multimodal model of 163
  natural 2, 159
Learning
  correlative Hebbian 39, 143
  delayed 129
  emotional 180
  instant Hebbian 38
  motor action sequences 128
LeDoux, J. 149
Linguistic description 160
Look-up table 43

MacDonald, J. W. 105
Machine
  cognition 137
  consciousness 185
  emotions 149
  perception 69
  self-consciousness 191
Malsburg, Von der, C. 188, 196
Marchetti, G. 174
Marr, D. 71
Massimi, M. 188
Match condition 41
Match/mismatch detection 41
Match/mismatch/novelty 41
McClelland, J. L. 11
McCulloch, W. S. 4
McGurk effect 105
McGurk, H. 105
Meaning
  associated 137, 160, 182
  categorical 169
  context 71
  intrinsic 137, 182, 186, 190, 191

Meaning of words
  horizontal grounding of 168
  vertical grounding of 165
Memory(ies) 140
  echoic 99, 105, 140, 162, 163
  iconic 140
  long-term 140, 142
  short-term 140
  skill 140
  visual 87
  working 140
Mental maps 113
Mental models 138, 144
Mind-body dualism 197
Mind-body problem 69, 186
Mirror neurons 162
Mismatch condition 41
Moore, B. C. J. 99
Morin, A. 193
Morton, H. ix
Motivation 149, 156
Motor action sequences 128
Motor control 117
Multimodal model of language 163
Multiple drafts theory 188

Nairne, J. S. 144
Natural language 2, 159
Neural networks
  approach 4
  artificial 4
  associative 17
Neuron(s)
  associative 19
  feedback 73
  Haikonen 19
  McCulloch-Pitts 17
  perceptron 18
  models 17
Newell, A. 3
Noë, A. 69
Noncomputable functions 42
Nonconscious acts 187
Nonconscious perception 186
Nonconscious processes 188
Novelty condition 41

O'Regan, J. K. 69
Ostension 165

Pain 150, 151, 152, 154, 198
Pandemonium 70
Parallel-to-serial transformation 56

Pattern recognition 18, 70, 138
Penrose, R. 5
Perception
 auditory 98
 haptic 82
 kinesthetic 81
 and recognition 70
 single feature 72
 of time 143
 visual 84
 with models 138
Perception/response feedback loop 72
Perceptron 18
Phoneme 54, 60, 103, 161
Physical symbol system 3
Pitch 102
Pitts, W. H. 4
Pixel map 85
Planning 145
Pleasure 150, 151, 152, 154, 198
Plutchik, R. 149
Port, R. F. 4
Prediction
 first order 79
 higher order 81
Predictive perception 79
Predictor, associative 57
Preprocessing
 auditory 99
 visual 85
Priming 74, 75, 76, 105
Pronouns 175
Proprioception 81
Pulse train signals 15

Qualia 190, 197
Quantum approach 5

Radvansky, G. 138, 173
Reactive actions 134, 135, 137
Reasoning 146
Recognition
 by components 71
 of object 89
 of speech 102
 of temporal sound pattern 101
Representation of
 graded values 14
 information 11
 significance 14
 words 161
Retina 84, 86, 91, 96, 138, 182

Robot
 cognitive 134
 liability of 198
Robotic arm 126
Rosenblatt, F. 18
Rosenfield, I. 144
Rumelhart, D. E. 11

Schachter, S. 149
Schneider, J. F. 193
Sejnowski, T. J. 22, 43
Self 144, 191, 192, 193, 194
Self-consciousness 191, 193
Self-evaluation 193
Self-image 83, 84, 193
Self-talk 193
Selfridge, O. G. 70
Sensorimotor coordination 117
Sensors 71
Sensory attention 137, 139, 155
Sentence comprehension 168
Sequences, timed 62, 63
Serial-to-parallel transformation 54
Shannon, C. E. 9, 160
Short-term memory 140
Siegrist, M. 193
Signal vectors 11
Significance
 emotional 149, 150
 representation of 14
Simon, H. A. 3
Singer, J. E. 149
Situation models 173
Sloman, A. ix, 3
Smith, L. B. 4
Soul 5, 197
Sound direction
 detectors 105
 perception 103
Sound pattern recognition 101
Spectrum analysis 99
Speech
 acquisition 161
 inner 2, 176, 180, 193, 196
 recognition 102
Steels, L. 193
Structuralism 70
Subjective feel 190, 197
Subjective feeling 149, 150
Subset interference 27, 36
Synapse 19, 23, 52, 86, 183
Synaptic circuit 20
Synaptic weight 19, 20

Syntax   10, 159, 160, 173
System reactions   149, 150, 154, 155, 180

Taylor, J. G.   ix, 69, 173
Template matching   70
Thelen, E.   4
Thinking machine   197
Threshold functions   20
Time   143
Timed sequences   63
Timing circuits   61
Torrance, S.   ix
Trehub, A.   ix
Treisman, A.   71
Tsakiris, M.   192
Turing test   196
Turing, A.   3, 196

Understanding of language   159

Valiant, L. G.   ix, 35
Value system   134, 150, 180, 198, 199

Visual   10, 84–100
   attention   86
   change detection   94
   distance estimation   92
   memory   87
   motion detection   95
   perception   84
   preprocessing   85
Vowel   102

Wallace, R.   4
Waveform   99
Weight matrix   23
Wiio, O.   160
Will
   free   194, 195
   own   195
Willed actions   156
Winner-Takes-All circuit   49
Wundt, W.   70

Zombie   149, 187
Zwaan, R.   138, 160, 173